The Age of Baroque

THE AGE OF BAROQUE

MICHAEL KITSON

Lecturer in the History of Art,
Courtauld Institute of Art, London

McGRAW-HILL BOOK COMPANY
NEW YORK · TORONTO

General Editors

BERNARD S. MYERS   **TREWIN COPPLESTONE**
New York   *London*

PREHISTORIC AND PRIMITIVE MAN
Dr Andreas Lommel, Director of the Museum of Ethnology, Munich

THE ANCIENT WORLD
Professor Giovanni Garbini, Institute of Near Eastern Studies, University of Rome

THE CLASSICAL WORLD
Dr Donald Strong, Assistant Keeper, Department of Greek and Roman Antiquities, British Museum, London

THE EARLY CHRISTIAN AND BYZANTINE WORLD
Professor Jean Lassus, Institute of Art and Archaeology, University of Paris

THE WORLD OF ISLAM
Dr Ernst J. Grube, Associate Curator in Charge, Islamic Department, Metropolitan Museum of Art, New York

THE ORIENTAL WORLD
Jeannine Auboyer, Keeper at the Musée Guimet, Paris
Dr Roger Goepper, Director of the Department of Oriental Art, State Museums, Berlin

THE MEDIEVAL WORLD
Peter Kidson, Conway Librarian, Courtauld Institute of Art, London

MAN AND THE RENAISSANCE
Andrew Martindale, Senior Lecturer in the School of Fine Arts, University of East Anglia

THE AGE OF BAROQUE
Michael Kitson, Lecturer in the History of Art, Courtauld Institute of Art, London

THE MODERN WORLD
Norbert Lynton, Head of the School of Art History and General Studies, Chelsea of Art, London

Library of Congress Catalog Card Number 65-21591

34900

© PAUL HAMLYN LIMITED 1966

PRINTED IN THE NETHERLANDS BY JOH. ENSCHEDÉ EN ZONEN
GRAFISCHE INRICHTING N.V.   HAARLEM

Previous pages: Mid-18th-century carved and gilt angel. (Austrian). Victoria and Albert Museum, London

List of Contents

Colour Plates

Rembrandt van Rijn: *A Girl Asleep*. *c*. 1655–56. Brush drawing in brown ink.
10 × 8 in. (24.5 × 20.3 cm.). British Museum, London.

Introduction

RENAISSANCE AND BAROQUE

The period known as the Baroque, which comprises broadly the 17th and 18th centuries in European art, can be looked on as a 'middle term' between the Renaissance and the modern age. In a sense it was the Renaissance enacted over again. Viewed from a point outside time it seems like a further phase of what had gone before, a renewal of creative activity on the lines that preceded it, a second revolution of the wheel.

Like all phases of art, the Baroque went through the familiar sequence of rise, climax and decline. But the resemblance to the Renaissance goes further than that. Baroque and Renaissance artists were faced with essentially the same kinds of task and worked for largely the same patrons: the court, the aristocracy and the Church. In this book we shall meet chiefly the same types of work of art: in architecture, churches and palaces; in sculpture, life-size marble statues made for altars, tombs and public squares or gardens; in painting, idealised figure compositions representing subjects from the Bible, classical history and mythology.

Like the Renaissance, the Baroque was dominated by the careers of great artists. The equivalents of Leonardo da Vinci, Michelangelo, Raphael, Titian and Dürer in the one period are Bernini, Poussin, Rubens, Rembrandt, Velasquez and Tiepolo in the other. More than this, both groups of artists shared many of the same basic aesthetic attitudes and looked back—the second partly through the eyes of the first—to the same source of inspiration: the Antique. Classical art, which meant chiefly the architecture and sculpture of ancient Rome (not Greece), is an almost constant factor in the background of the works of art to be discussed in this book. Its relationship to the Baroque is symbolised in the etching by Piranesi illustrated as figure 1.

Yet, as we watch the wheel turn and the sequence of styles come round again, we notice, as we should expect, important differences as well as similarities. In the first place, almost everything is pitched in a higher key, the tone is more florid and colourful, the textures are richer, there is more decoration, more light and shade, apparently less control. The lyrical refinement characteristic of the Renaissance is not often found; where there is restraint, it is apt to appear as severity. There is also sometimes deliberate striving after effect.

In the second place, the baroque period was accompanied by an expansion of art on all fronts, in its geographical limits, in its patrons and in its categories. It was not so much that the centre changed from any of these points of view (although Paris replaced Rome as the artistic capital of Europe almost half-way through the period) as that the fringes became more important. The spread of art can be seen in the rise or revival of national schools in France, Spain, Holland, England and Central Europe. These all contributed far more to the Baroque than they had to the

Renaissance, which had been dominated by Italian art, although the Baroque itself also had Italian origins. This upsurge of creativity north of the Alps reflected the growth of national states and a shift in economic power, which continued throughout the period, from the Mediterranean lands to the countries bordering the Atlantic, especially England and France. Moreover the Baroque was carried overseas by Catholic missionaries to Latin America and the Far East; colonial settlement also brought English architectural styles to the eastern seabord of North America in the 18th century.

In patronage, the middle classes began to play an active though not yet a decisive part, except in Protestant Holland where there was no official court or aristocracy and no demand for paintings from the Church. But the clearest way in which art expanded was in the growth of the so-called 'lesser categories'. This is especially visible in painting. The lesser categories included portraiture, landscape, *genre* painting (the representation of subjects from everyday life) and still-life painting. For the first time, apart from portraiture, these now took a significant place alongside the traditionally superior categories of painting whose subjects were drawn from the Bible, classical history and mythology. In architecture, town houses and town-planning assumed the same role; in sculpture this role was taken by portrait busts. In the 18th century, interior decoration and furniture entered the mainstream of stylistic developments for the first time, and one entirely new art was created during this period: porcelain.

The third basic change brought about by the Baroque lay in the increasing complexity of stylistic trends. It is true that the Renaissance was by no means a single, unified movement. It reached its climax in different places at different times, and its long, extremely varied final phase —Mannerism—in some ways marked a reaction against the period of its highest and most typical achievements. But the Renaissance was still far more homogeneous than the Baroque. In fact, the word 'Baroque' is not properly adequate to describe the art of the 17th and 18th centuries except in a very general sense. Although it was the dominant style, which coloured all others in varying degrees, in the strict sense it was only one style among several others which developed simultaneously, as must now be explained.

THE MEANING OF STYLISTIC TERMS AND
THE THEME OF THIS BOOK

Nowadays 'Baroque' is used as a stylistic term (as distinct from a period label, as in the title of this book) to describe the art that first arose in Italy shortly before 1600, flourished there until the mid-18th century and spread particularly to Flanders, Germany, Central Europe (i.e. Austria, Bohemia and Poland), Spain and the Spanish colonies overseas, although it also produced echoes of varying intensity in the art of all other European countries. Alongside the Baroque a classical movement grew up in

1. **Giovanni Battista Piranesi.** *The Arch of Septimius Severus with the Church of SS. Luca e Martina.* 1759? Etching. 14 × 22½ in. (35.5 × 57.5 cm.). This etching, from the series, *Le Vedute di Roma*, is reproduced here to underline the importance of Rome in this period and to symbolise the relationship of baroque art to the Antique. A 17th-century church—Pietro da Cortona's SS. Luca e Martina (1635–50)—faces classical ruins in the ancient Roman Forum. Rome was the artistic capital of Europe in the first three-quarters of the 17th century and again assumed a key role in the age of Neo-Classicism.

partial opposition to it, finding its particular home in France. There was also a third style, Realism, associated with such artists as Caravaggio and many Dutch painters.

All three styles continued, with modifications, into the 18th century. But the 18th century also saw the creation of a new style, the Rococo, whose relationship to the Baroque has some similarities with the relationship of Mannerism to the Renaissance; that is, it was partly an adaptation of the earlier style and partly a reaction against it. By the 1760's, classicism had merged into Neo-Classicism and had become the characteristic style of the age. This broke the dominance of the Baroque and created a link with the modern period.

It also broke the cycle which had begun just before 1600, so preventing the wheel from completing its revolution; that is to say, although both the baroque and rococo styles declined, the period ended on a rising not a falling note. In a sense this was also true of Mannerism, although the break between it and the Baroque was sharper, for many of the developments that flourished in the 17th century were conceived in the second half of the 16th century, just as Neo-Classicism was not only a backward-looking movement but also one that laid the foundations of modern attitudes to art.

The tracing of these stylistic movements, together with a discussion of their characteristics, forms the theme of this book. In general history, the period corresponds to the heyday of the *ancien régime*, the power struggle between European states and such intellectual currents as the later stages of the Counter-Reformation, the Enlightenment, and the beginnings of modern science. Soon after the period opened the Thirty Years' War (1618–48) began; its end was marked by the outbreak of the French Revolution (1789).

Some explanation should perhaps be given as to why styles rather than national schools have been chosen as the units into which this book is divided. It might be said that such an approach makes artists appear to conform to trends which have only been defined in retrospect; moreover, the temptation is to endow these trends with a will of their own (making the Baroque 'do' this, the Rococo 'do' the other), whereas in fact they have no existence apart from the works of art they describe. But the styles detected by modern art historians are related to issues which were actually felt by at least some artists at the time. Against a background of the Renaissance as well as classical antiquity, these artists were among the first in history able to exercise a conscious stylistic choice. The very number of schools and masterpieces created by the Baroque militates against too factual an approach in so short a book as this. Furthermore the characteristics which cut across national frontiers and the boundaries between the arts are of exceptional interest in this period, making a deeper understanding of its artistic achievements possible than would have been the case with a straightforward chronological and geographical survey.

THE ORIGIN OF 'BAROQUE' AND 'ROCOCO'

Like most art-historical terms before the late 19th century, 'Baroque' and 'Rococo' were both introduced in retrospect and were first employed in a spirit of abuse. 'Baroque' seems to have been derived either from a Portuguese word, *barroco*, meaning an irregularly shaped pearl, or, as some historians assert, from an Italian term, *baroco*, a stumbling block in medieval scholastic logic. In either event, the word had acquired currency in a metaphorical sense in Italian and French by the 16th century, when it meant any contorted idea or tortuous, involved process of thought.

Its application to art did not begin until the second half of the 18th century, during the ascendancy of Neo-Classicism. It was then, and up to a point still is, defined in terms of its opposition to classical values. In 1771 the *Dictionnaire de Travaux* gave one meaning of Baroque (the other referred to its use as a synonym for eccentric or bizarre) as 'in painting, a picture or figure... in which the rules of proportion are not observed and everything is represented according to the artist's whim'. Sculptors and architects who offended against classical rules were even more severely attacked than painters. The Italian critic, Milizia, influenced by the great neo-classical theorist, Winckelmann, wrote in 1797: 'Baroque is the ultimate in the bizarre; it is the ridiculous carried to extremes. Borromini went delirious, but in the sacristy of St Peter's, Guarini, Pozzi and Marchione went Baroque.'

In the 19th century, the word continued to be used chiefly of certain aspects of Italian 17th-century architecture, although the hostility towards the style it referred to spread to the other arts and also to the art of other countries. It was not until the publication of Heinrich Wölfflin's *Renaissance und Barock* in 1888 that 'Baroque' was neutralised for art-historical purposes, although Wölfflin applied it to an earlier period—the hundred years from about 1530 to 1630—than is now usual. Even in Wölfflin's time it was only in Germany and only among scholars that baroque art was considered respectable; elsewhere it was still regarded as a debased continuation of the art of the Renaissance. In England and America popular prejudice against the Baroque continued almost until the last war. Since then it has won general acceptance, partly owing to the contemporary readiness to consider any style on its merits, rather than judge it in advance by abstract aesthetic standards, and partly because of the appeal of the sheer energy, daring and inventiveness of the Baroque to modern taste.

2. *The Interior of the Gesù, Rome*. Architecture by **Giacomo Vignola,** begun 1568. The baroque use of rich materials and ornate decoration as a means both of glorifying God and appealing to the worshipper's emotions is shown here in the mother church of the Jesuit Order in Rome. At first the interior was comparatively bare, and it only received its present form in the late 17th century. The frescoes in the apse, pendentives, dome and nave (fig. 29) were painted from 1672–83 by Gaullì.

The origin of 'Rococo' is rather the same. In its purest form the Rococo was a fanciful decorative system devised in France about 1700 for the treatment of interiors and ornamental objects generally, as an escape from the pompous, monumental style typified by Louis XIV's Versailles. In these fields—ornament and decoration—it reached its zenith in Paris in the 1730s, spread to other arts and other countries, particularly Germany, and was later associated retrospectively with the age of Louis XV. *Le style Louis XV*—inaccurately—or *le style Pompadour* —even less accurately—as it came to be known, was first called 'Rococo' by the early Romantic artist and pupil of David, Maurice Quaï, in 1796–97. It was probably derived from *rocaille*, an adjective referring to the shells and bits of rock used in 16th-century decoration, and became current in Paris studios as a nickname for the style against which

3. **El Greco.** *The Immaculate Conception. c.* 1613. Oil on canvas. 128 × 65¾ in. (323 × 167 cm.). Museum of S. Vicente, Toledo. Although El Greco's style is usually placed under the heading of Mannerism, his late work is in many ways baroque in feeling. He was among the first artists to develop the continuous upward-soaring movement typical of the Baroque and to convey emotion by means of fluttering draperies and ecstatic expressions and gestures. The Immaculate Conception of the Virgin Mary was one of the devotional subjects brought into prominence by the Council of Trent.

4. **Peter Paul Rubens.** *Apotheosis of James I* (detail). 1629. Oil on panel. 37½ × 25 in. (95 × 63.1 cm.). Mrs Humphrey Brand, Glynde Place, Sussex. This was probably Rubens's first sketch for the ceiling of the Banqueting House, Whitehall, which was commissioned by Charles I in honour of his father and the Royal House of Stuart. The canvases were completed in Antwerp and set up in place in 1635. It shows the use of art as propaganda on behalf of a secular monarch, corresponding to the use of religious art for similar purposes in Catholic churches. Rubens's sketch, in shades of silver-grey and honey-yellow, is a superlative example of his brushwork.

artists were by then in full revolt. In effect, what the Baroque was for Winckelmann and the Italian neo-classicists, the Rococo was for neo-classicists in France. The rehabilitation of French rococo painting, decoration and furniture began over a century ago, much earlier than the Baroque, but it is only quite recently that the Rococo has been critically assessed.

THE INFLUENCE OF RELIGIOUS AND
PHILOSOPHICAL IDEAS

Very broadly, the 17th century was an age of authority, the 18th an age of scepticism. This should at once be qualified by pointing out that the spirit of scientific enquiry which eventually undermined authority first appeared very early in the 17th century, while the authoritarianism that determined the social structure at the beginning of the

period persisted almost unchanged until the French Revolution. Nevertheless, the association of the first of these two centuries with hierarchy, precedence, degree and the maintenance of authority, and the second with liberty, scepticism, rationalism and the challenge to authority, generally holds good.

The 17th-century deference to authority can be seen, for example, in the doctrine of the divine right of kings, the Church's claim to interpret all knowledge and experience in the light of its teaching, and the way in which artists justified their theoretical principles by appealing to classical sources. None of these things was new but they all became more pronounced in this period. In secular affairs, power was concentrated more than ever in the hands of the ruler or his first minister: for example, Pope Urban VIII, Cardinal Richelieu, Philip IV of Spain, Louis XIV of

5. **Guido Reni.** *Ecce Homo. c.* 1640. Oil on canvas. $44\frac{1}{2} \times 37\frac{1}{2}$ in. (113×95 cm.). Fitzwilliam Museum, Cambridge. Like El Greco's *Immaculate Conception* and Lebrun's head representing *Terror* (fig. 6), this is an example of the all-absorbing subject for 17th-century artists of emotional expression, which reflected the contemporary interest among theologians and philosophers in the psychology of the soul. When the treatment of emotion is as direct and charged with pathos as it is here it is apt to seem distasteful nowadays, but it is highly characteristic of baroque art and cannot be dismissed as insincere.

6. **Charles Lebrun.** *The Expression of Terror.* 1668? Pen and black ink over a sketch in black chalk. $7\frac{3}{4} \times 10$ in. (19.5×25.5 cm.). Louvre, Paris. The concern with expressing emotion originated in the Renaissance and was taken up in the 17th century by both classical and baroque artists. Lebrun, director of the French Academy of Painting and Sculpture, 1663–90, attempted to reduce the problem to rules, and this diagrammatic head, showing the way the facial muscles contract under the stress of terror, was one of a series drawn by him to illustrate his lecture on 'The Expression of the Passions'. The lecture, probably first given in 1668, was first printed in 1696, and was widely influential.

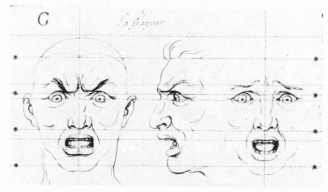

France, Colbert, Charles I of England, and Cromwell. In religion (taking the Catholic view), the tide of Protestantism was halted—in some places turned back—and the Catholic Church felt a renewed sense of confidence in its mission, contrasting with its anxious, defensive stance in the previous century. The religious orders founded during the first phase of the Counter-Reformation also attained their greatest power and influence in this period. This was particularly true of the best organised and most militant among them, the Society of Jesus, founded in 1540.

The effects of the Reformation on the arts were largely negative, since religious paintings and sculpture were banned from Protestant churches, but the Counter-Reformation, through its doctrinal instrument, the Council of Trent (1545–63), had far-reaching consequences. The representation of heretical ideas and indecent or irrelevant matter was forbidden; pictures and statues of the Virgin Mary, the martyrs and saints, particularly in states of ecstasy or meditation, increased; and religious art was encouraged in churches provided it gave instruction in the Faith (in an almost medieval sense) and was conducive to piety.

From the last of these proposals it was a short step to the use of art as a means of propaganda—an idea that was also seized on by secular rulers for their own ends. Of its nature, propaganda is designed to reach the mind of an audience by appealing to its emotions, and there is a predictable relationship between the more emotional and rhetorical

—in general the more baroque—forms of art in the 17th century and the patronage of courts and the Catholic Church. On the whole, baroque art in the strict sense is associated more with Catholic than Protestant countries (see endpaper map), more with countries particularly loyal to the Pope (e.g. Spain) than those which, while still Catholic, followed a comparatively independent religious policy (e.g. France), and more with countries that remained feudal and rural in their political and social organisation than those that became more industrialised and 'advanced' (this is shown by the continuance of the Baroque in the 18th century in countries like Spain and Southern Germany, whereas France and England turned to the more refined and worldly styles of the Rococo and Neo-Classicism).

This is not to say that the Baroque was consciously evolved by Catholic, monarchical patrons as a deliberate instrument of policy or that it was an inevitable outcome of their societies; still less is it to suggest, as was once the fashion, that the Baroque was the invention of the Jesuits. In fact the Baroque, like most other styles, was invented by artists and found expression as beautifully and effectively in small, private works as in large, public ones, although the latter are its most typical manifestations. Moreover, the Church's attitude towards art, particularly the attitude of the religious orders, was still comparatively austere in the first half of the period, though as much perhaps for reasons of economy as from principle; the rich baroque altars and

7. **Charles Nicolas Cochin the Younger.** *The Drawing School in the Academy.* 1763. Engraving by B.L. Prévost. 4 × 8 in. (10.2 × 20.4 cm.). Although dating from the second half of the 18th century, this engraving represents the curriculum introduced in the French Academy in the 1660s—a curriculum imitated with minor variations by academies all over Europe. The young students are seen copying the work of other masters, while the older ones are drawing from life and from casts after the Antique.

decorations that fill Catholic churches today were only introduced in the second half of the 17th and in the 18th centuries.

Another point that needs clarification is the degree to which fluttering draperies and exaggerated facial expressions and gestures in baroque paintings and statues were intended to stir the worshipper into sharing the emotions of the person or persons represented. A test case of this is Bernini's statue of the *Ecstasy of St Teresa*, made for the Cornaro Chapel in Sta Maria della Vittoria, Rome, in 1645–48, the subject of which is the saint's vision of the love of God, which she experienced in the form of a flaming dart thrust into her by an angel. There is no doubt that this statue is a highly charged and sensational work, but Bernini was probably at least as much concerned with *representing* the emotion described by the saint as with *inducing* a corresponding emotion in the observer. In fact Catholic theologians, especially the Jesuits to whom Bernini was closely attached, discouraged too much indulgence in feeling during meditation or prayer, as being likely to lead to sin when the feeling had worn off. What contemporary theologians were very interested in, however, was the psychology of the soul. The state of mind of the saint, the martyr and Christ himself were minutely speculated upon and analysed. Concern with the analysis and vivid representation of emotion played a correspondingly important part in 17th-century aesthetic theory. In so far as the observer was intended to identify his own emotions with

the subject-matter, the explanation is likely to be found in aesthetic theory rather than theology. Some theoretical writers drew a parallel between the visual arts and rhetoric—the art of persuasion—and evolved a primitive doctrine of empathy, although this was not new in the 17th century.

AESTHETIC DOCTRINES OF THE PERIOD

The 17th-century deference to authority found direct expression in the arts, first, in the constant quotation of classical sources in both practice and theory, and, second, in the doctrine of the hierarchy of categories, according to which large-scale figure paintings with subjects from the Bible, classical history and mythology were considered intrinsically more important than portraits, landscapes, *genre* scenes, and so on. It may also be said that a baroque church façade or figure composition was modelled on a hierarchical system, with a climax at or near the top and with each part related to the others is descending order of importance, the lesser always enclosed within the greater, down to the smallest details. Aesthetic theory revolved round these 'higher' categories, and it was not until the 18th century that writers began to go more deeply and intelligently into the aesthetic problems of portraiture, landscape and *genre* painting.

The central aesthetic doctrine of the period was the belief that painting and sculpture were modes of imitating ideal nature—ideal not actual nature, because it was held,

8. **Roland Fréart de Chambray.** *Plate from the 'Parallèle de l'architecture antique et la moderne'* (detail from the 2nd ed., 1689). 1st ed., Paris, 1650. 11 × 7¼ in. (28 × 18.4 cm.). Like the interest in expression, the systematic study of the five classical orders of architecture (Tuscan, Doric, Ionic, Corinthian and Composite) began in the Renaissance and was continued in the 17th and 18th centuries, especially in France. Fréart de Chambray was one of the first to advocate strict adherence to the rules laid down by the Roman writer, Vitruvius, and he used engravings after measured drawings from classical remains to support his arguments. The plate shown here illustrates the Doric Order from the Baths of Diocletian, Rome.

Aux Termes de Diocletian a Rome

following Plato and Aristotle, that actual nature was always imperfect in some way. It was the business of the artist to discover and represent the perfect or ideal forms of things, his guide to the ideal being the sculpture of the ancient Greeks and Romans and, in modern times, the paintings of Raphael. Importance was also attached to decorum, that is, the principle that every form in a picture or piece of sculpture should be represented in a style appropriate to its subject-matter. The same was true in architecture. The correct classical 'order'—Tuscan, Doric, Ionic, Corinthian or Composite—had to be used for the appropriate type of building, and if more than one order was employed they had to be arranged above each other in the sequence sanctioned by classical precedent and justified by the ancient Roman writer on architecture, Vitruvius.

All these doctrines were expounded repeatedly in theoretical treatises, particularly in the second half of the century, often in association with academies. The most important academies were the Roman Academy of St Luke, founded in 1593, and the French *Académie royale de peinture et de sculpture*, founded in 1648 but not fully active until the 1660s. Academies were another means of enforcing authority in the arts, though they were also concerned with improving the status of the artist and in the 18th century with organising exhibitions. The most representative theoretical treatise was *The Idea of the Painter, Sculptor and Architect* by G. P. Bellori, published in Rome in 1672.

Academies and theoretical writings both contained a classical as opposed to a baroque bias, but the principal aesthetic doctrines, all of which originated in earlier periods and were only elaborated and codified in the 17th century, were subscribed to by baroque and classical artists alike. One of the few subjects of debate seems to have been the relative importance of form, which classical artists stressed, and colour, which was emphasised by artists of the Baroque. Otherwise the distinguishing features of the Baroque, such as the dynamic treatment of space, movement and

light and shade, fell outside the scope of theory. It is a curious but striking fact that the Baroque had no specific aesthetic programme of its own. This was even more true of the Rococo, which was a consciously anti-theoretical movement.

The debate over form and colour came to a head in the French Academy at the end of the 17th century, resulting in the victory of the 'colourists' and the substitution of Rubens for Raphael and Poussin as the hero of contemporary artists. But this did not mean the replacement of one orthodoxy by another; it meant, rather, the weakening of orthodoxy altogether. A more liberal attitude towards the imitation of antiquity was also adopted following the 'Quarrel of the Ancients and the Moderns'. In other words, the challenge to authority which was now beginning in philosophy was finding expression in the visual arts. The high seriousness which had informed 17th-century figure painting and sculpture went out in the first half of the 18th century, and was replaced by a frank delight in purely visual and decorative effects. The aesthetic qualities admired were 'variety', 'charm' and 'grace'. There was a new appreciation of the evocativeness of sketches and drawings, precisely because they were unfinished in the conventional sense. Works of art were now less often judged by reference to absolute rules and standards than by intuition.

But if scepticism and the general distrust of all dogmas and systems taken from sources outside sense experience found a perfect reflection in the Rococo, that other component of 18th-century thought—rationalism—led to a classical revival and eventually to Neo-Classicism. Neo-Classicism, in sharp contrast to the Rococo, was deeply inspired by theory. Perhaps it is the only movement in the history of art to have been brought into being by critics, philosophers and connoisseurs rather than artists. However, its style is so closely bound up with the ideas underlying it that it will be best to postpone any discussion of those ideas until the appropriate chapter.

The Baroque

It will be best to begin this account of baroque art in the strict sense with a point that has been touched on already in discussing religious propaganda and rhetoric, namely the way in which baroque art appeals to the mind through the emotions. Of course, all art appeals in varying proportions to both the emotions and the mind. But the Baroque makes use of emotional appeal as a means of reaching the mind in a special way. It goes out to meet the spectator's emotional susceptibilities; it is 'spectator-orientated' to a greater extent than any other style. Unlike the diffuse, tortuous style of some forms of Mannerism, it is visually easy to read.

To give an example, the design of baroque ceiling decorations is calculated, having regard to the shape, size and lighting of the room, to make it as convenient as possible for the spectator to view the ceiling from the position or positions he would normally take up. This factor had scarcely been considered in the previous period, when ceilings were designed to be seen either from one viewpoint only or from a variety of conflicting viewpoints. Mannerist painters were also apt to treat the ceiling on either too large or too small a scale for the height of the room, filling low ceilings with very large forms and high ceilings with very small ones. The characteristic of baroque ceilings is that they are optically 'just right'.

To give another example, figures in easel and wall paintings tend to be concentrated in the foreground plane. Caravaggio, one of the revolutionary painters of the new age, had begun to do this even before 1600, although in many ways he was still a pre-baroque artist. In almost all his pictures the background is shut off by a wall of darkness, leaving only a shallow stage in front for the figures to move in. Mood, action and the physical reality of bodies are thus insistently pressed on our attention, involving us in the imagined world of the artist's creation. This device is repeatedly used, not only in Caravaggio's art but also —in fact slightly earlier—in that of Ludovico and Annibale Carracci, and later in the work of Guido Reni, Pietro da Cortona, Rubens, van Dyck, Rembrandt (though not consistently) and many others. The darkness lightens in these later examples and is characteristically varied with half-lights and reflected lights or is replaced altogether by a pale grey neutral background or a landscape and clear sky. But the figure or figures remain compellingly 'there'—alert, expressive and watchful. Sometimes they gaze at or gesture directly towards us, as if to penetrate the psychological barrier between their world and ours.

The psychological impact resulting from baroque effects of close-up can also, paradoxically, be produced by the opposite means—emphasis on the extension of space in depth. This was chiefly used in the high and late baroque periods (i.e. from the 1630s onwards) and is most characteristic of the last third of the 17th century. In painting it occurs particularly in illusionistic ceiling decorations, but the spatial depth of Claude's landscapes, which are scarcely

baroque in any other sense, are another example. Architectural examples of this use of extended space are the staircase and the vista, both of which acquired a new importance in the baroque period. With paintings the spectator's feeling of being drawn into the space is naturally subjective, but with buildings it can and should be translated into action. Baroque buildings must, that is to say, be walked round and through and be studied, as it were, on the move if their true quality is to be realised. The problems of baroque space will be discussed in more detail later.

ILLUSIONISM

A further, related means of heightening the emotional appeal of a work of art was through the use of illusionism. Illusionism was far from being an invention of the baroque period, nor are all baroque works of art illusionistic, but it now became more common and more convincing than ever before. In one sense illusionism only represented the extension of an ambition that artists had first had in antiquity and revived in the early Renaissance—the ambition of complete realism. Vasari often praises a work for being so lifelike as almost to deceive the eye. Admittedly, by the High Renaissance mere optical deception was no longer considered enough, and the literal copying of nature, without selection or idealisation, was despised by 17th-century critics. Nevertheless, the vivid and convincing rendering of ideal nature was still regarded as one of the painter's and sculptor's primary tasks. Even in the 17th century the creation of a lifelike image out of the inanimate materials available to the artist contained an element of the miraculous, and the technical skill required to produce such an image was highly prized. 'It is an imitation made with lines and colours on a flat surface of everything under the sun' said (or rather quoted) Poussin, and this was still the basic definition of painting. It remained so for conservative critics until the 19th century.

However, there is a distinction to be made here between illusionism and realism. Realism meant the faithful rendering of the outward appearance of objects, whether such objects belonged to the actual world or were only present to the imagination; idealisation was therefore excluded. Illusionism, on the other hand, was not only consistent with idealisation and the imaginative representation of supernatural events and experiences, but was often combined with them.

Typical examples of illusionism are Bernini's *St Teresa*, Pozzo's ceiling of S. Ignazio in Rome and Egid Quirin Asam's *St George* altar in the monastery church at Weltenburg. None of these depict situations or events that could have been seen with the natural eye, yet they brilliantly project into visual terms what might have been seen by the eye of the imagination. They appear to be so convincing that 'they might almost be there'.

A further use of illusionism which these examples show consisted in overcoming the natural limitations of materials; marble was made to look like hair or cloth, gilded

20,93
32,8
2,40
45
1,42
6,47
1
32
9

16

9. **Egid Quirin Asam.** *High Altar with St George killing the Dragon.* 1721. Altar figures in gold and silver gilt stucco, the twisted columns in marble, the rest in wood and stucco, partly painted and gilt. Monastery Church, Weltenburg, Bavaria. Besides its spectacular illusionism (see text), this altar is characteristic of the Baroque in its use of richly coloured materials, heavy architectural forms and broken pediments and twisted columns. The church was built in 1716–18 by Egid Quirin's brother, Cosmas Damian Asam, who also designed the fresco of the *Immaculate Conception* painted on the back wall of the brightly lit space behind the altar.

metal like rays of light, or a painted picture-frame like a real one. Nor was illusionism confined to painting and sculpture. In architecture the design of a building did not necessarily correspond to its structure, and façades and *38* screen walls were freely used to mask an irregular or un-pleasing feature behind them. The modern doctrine of functionalism, which was founded in the age of Neo-Classicism, had no place in the Baroque.

Without doubt the most versatile master of illusionism was Bernini. His command of it is as apparent in small things as in great. During his visit to Paris in 1665, when he executed the marble portrait bust of Louis XIV, Chantelou recorded his observation: ' "Sometimes in order *22* to imitate the model well it is necessary to introduce in a marble portrait something that is not found in the model." This seems to be a paradox, but he explained it thus: "in order to represent the darkness that some people have around the eye, it is necessary to deepen the marble in that place where it is dark in order to represent the effect of that colour and thus make up by skill, as it were, the imperfec-tion of the art of sculpture, which is unable to give colour to objects." ' Of an earlier work by Bernini, the *David* in the *10* Borghese Gallery, the sculptor's biographer wrote: 'The magnificent head of this figure (in which he portrayed his own features), the vigorous down-drawn, knitted eyebrows, the fierce, fixed eyes, the upper lip biting the lower, marvel-lously express the righteous anger of the young Israelite taking aim with his sling at the forehead of the giant Philistine. The same resoluteness, spirit and strength are found in every part of the body, *which needs only movement to be alive*' (present writer's italics).

Another way in which illusionism enhanced the vividness of a work of art was through the use of devices designed to associate it with the real world of the spectator. In this sense the *David*, executed in 1623, was one of the first true ba-roque statues, for its whole stance and gaze suggest move-ment beyond the limits of the sculpture itself. Its focus is not within the work of art but outside it, in the spectator's space. The same effect can be seen, though not carried to such lengths, in Rembrandt's painting, the *Night Watch*, in *7* which the leading figures appear to be marching out of the picture towards us. It occurs again in the type of equestrian portrait in which the horse and rider are shown head-on. This motive was actually invented by El Greco in the 1590s

(Continued on page 33)

1. **Gianlorenzo Bernini.** *The Ecstasy of St Teresa.* 1645–52. Marble. Life-size figures. Cornaro Chapel, Sta Maria della Vittoria, Rome. This is one of the touchstones of the Baroque, combining as it does sculpture, architecture and painting—or, at least, an effect analogous to painting—and projecting a powerful sense of illusion. Nature is brought into the work of art by the use of light from a concealed source, filtered through yellow glass, to re-inforce the symbolic light represented by metal rays. The artist's object is to create a convincing visual equivalent of St Teresa's own account of her mystical experience.

2. **Narciso Tomé.** *The Transparente.*
Completed 1732. Marble and other
materials. Toledo Cathedral, Spain. The
highly charged effects shown in the
previous illustration are here carried to
their ultimate and most dazzling conclusion. The altar is a setting for the
Blessed Sacrament, seen through a glass-
fronted receptacle (hence 'Transparente').
It is lit from above and behind the spec-

tator's head through a high window,
surrounded by sculptured figures of
Christ, prophets and saints, and let into
the Gothic vaulting of the ambulatory.
3. (above). **Peter Paul Rubens.** *The
Adoration of the Kings.* 1634. Oil on canvas.
129¼ × 97¼ in. (328 × 247 cm.). King's
College Chapel, Cambridge; reproduced
by courtesy of the Provost and Fellows.
Rubens's *Adoration of the Kings*, painted for

the Convent of the Dames Blanches at
Malines, exemplifies the power and
eloquence of the baroque altarpiece at
its most magnificent. The three great
figures of the Kings enact the offering up
of worldly authority to the Holy Child.
The artist's style, now past its most
aggressive phase, combines his usual
energy with a new note of tenderness and
composure.

4. *St Peter's, Rome.* Proto-, Early and High Baroque are here combined in the dome (by **Michelangelo,** 1547), façade (by **Maderna,** 1606) and colonnade (by **Bernini,** 1657) of St Peter's—the dates are those of the designs. The three parts link the ages. The first is summed up in the majesty and grace of the dome, a resolution in pure architectural terms of mass and soaring energy possible in quite this form only in the mid-16th century. The second is typified by the grandiose nave and façade added to the central crossing of the Basilica for liturgical reasons, a step characteristic of the practical, early 17th-century phase of the Counter-Reformation. Representing the third age, the ceremonial repetition of the columns enclosing the oval piazza and symbolising the all-embracing arms of the Church are typical both of High Baroque spatial concepts and of the period's revived interest in symbolism.

5. **Jakob Prandtauer.** *The Monastery of Melk,* Lower Austria. Begun 1702. Standing on a high rock overlooking the Danube and fully exploiting the visual possibilities of its position, Melk is one of the most spectacular buildings of Central European Late Baroque. In feeling it contrasts markedly, and characteristically, with St Peter's. Symbolism, intellectual content and the use of a strict vocabulary of architectural form are subordinated to a single end, the creation of an over-whelming scenic effect.

6. **Diego Velasquez.** *Philip IV of Spain.*
1634–35. Oil on canvas. 118½ × 124 in.
(301 × 314 cm.). Prado, Madrid. Painted
for the Hall of the Realms in the Palace of
Buen Retiro, Velasquez's *Philip IV* belongs
to a tradition of equestrian portraiture
going back to Titian's portrait of the
king's great-grandfather, the *Emperor
Charles V*, (already in the Spanish Royal
Collection), inviting a comparison not
only between the two pictures but also, as
contemporaries would have understood,
between the two monarchs. Philip is
presented as the more spirited figure,
riding alone across open country, not, like
the Emperor, as the commander of troops
in battle. The artist has created this
unforgettable image of Absolute
Monarchy—stiff, unyielding and
remote—without trappings, by pictorial
means alone.

7. **Rembrandt van Rijn.** *The Night Watch.* 1642. Oil on canvas. 152½ × 197½ in. (387 × 502 cm.). Rijksmuseum, Amsterdam. In fact, this represents a parade of the militia company of Captain Frans Banning Cocq and is a daylight scene, not a 'night' watch; but *The Night Watch* is the traditional and unpedantic title (acquired in the 19th century when the picture was covered with brown varnish), which should be retained. The civic group-portrait was the main type of public commission open to 17th-century Dutch artists. The finely dressed, dramatically lit figures, with drums beating and banner unfurled, are formed into a pattern of baroque complexity, their leaders 'walking out' of the picture towards the spectator.

8. (following pages). **Giambattista Tiepolo.** *Apollo conducting Beatrice of Burgundy to Frederick Barbarossa.* 1751–52. Fresco. Ceiling of the *Kaisersaal*, Residenz, Würzburg. At the height of his career, Tiepolo was called to Würzburg to decorate the dining-room of the Prince-Bishop, Karl Philipp von Greiffenklau; the room itself, an eight-sided rococo *salon* with multi-coloured marble columns and elaborate stucco-work, was designed by Balthasar Neumann. Tiepolo's illusionistic ceilings are the culmination of a great tradition; but although, as in the case of his baroque predecessors (see plate 46), the theme—the 12th-century episode of the Betrothal of Beatrice of Burgundy to the Emperor Frederick Barbarossa—is grave, Tiepolo's treatment is all lightness and rococo. The actors are in 16th-century fancy-dress *à la Veronese*; there is occasionally a playful element of genre, and disembodied *putti* flit in and out of Antonio Bossi's glittering stucco arabesques which frame the ceiling. It is painting not for instruction, but for decoration and charm.

9. (left). **Gianlorenzo Bernini.** *The Angel with the Superscription.* 1668. Marble. Over life-size. S. Andrea delle Fratte, Rome. Deriving from a series of angels holding the instruments of Christ's Passion made for the Ponte Sant'Angelo, this figure and one other were kept indoors on the Pope's orders, replicas being substituted on the bridge to save them from weathering. In style it shows an even more extreme form of Baroque than the angel in *The Ecstasy of Sta Teresa* (plate 1). Initially, the pose was derived from a classical statue, but in developing the design, Bernini took it far from classical norms of style. The virtuosity of the cutting, irregularity of the outlines and the near-affectation of the expression and movements of the hands were a starting point for much 18th-century German sculpture.

10. (above). **Nicolas Poussin.** *Confirmation.* 1645. Oil on canvas. 46 × 70 in. (117 × 178 cm.). Collection: the Duke of Sutherland, Mertoun; on loan to the National Gallery of Scotland, Edinburgh. From Poussin's second series of *Seven Sacraments*, painted for Fréart de Chantelou, this is 17th-century classicism at its purest and best. The quality it shares with the Baroque of Bernini, Rubens or Rembrandt is its high seriousness; stylistically it could hardly be more different. With Poussin, the intellectual discipline which he imposed on the free play of imagination resulted in its own special kind of beauty.

11. *Table. c.* 1680. Carved and gilt wood. Hotel Lauzun, Paris. This massive, ornately designed side-table is typical of French furniture of the reign of Louis XIV; it would have fitted well with the early interiors at Versailles (see next page), where the furnishings, conceived as part of a unified decorative ensemble, were more unrestrainedly baroque than the architecture. The volutes, acanthus leaves and garlands are treated naturalistically, in keeping with the taste of the period for illusionism.

13. (above). **François Cuvilliés.** *Hall of Mirrors*, The Amalienburg, Schloss Nymphenburg, Munich. 1734–39. The *Maison de Plaisance* of the Amalienburg was built by Cuvilliés for the Bavarian Elector, Karl Albrecht, as a counterpart to a pseudo-Gothic 'hermitage' in his palace park. The stucco is by J. B. Zimmermann, the carving by J. Dietrich. The pavilion is one of the earliest and most perfect examples of German Rococo, although its Flemish-born architect was trained in France and its exterior fulfilled French architectural theories of the period. The reduction of a hall of mirrors *à la Versailles* to the scale of a pleasure pavilion is matched by the light and sculptural forms of the decoration and is wholly in the spirit of Rococo.

12. (left). *Salon de Vénus*, Versailles. *c.* 1671 onwards. This room, decorated for Louis XIV under the direction of **Lebrun,** is one of six arranged *en suite* forming the *Grand appartement du réception du Roi*; they are the earliest in the Château that survive. The iconography of the rooms was based on the planets, the last room, containing the throne, being dedicated to the Sun-God, Apollo, with whom the King identified himself. The subject of the ceiling painting in the *Salon de Vénus* is the influence of love on kings. The statue in the niche of Louis XIV is by Jean Warin.

The capitals and over-doors are in gilded copper, the walls in coloured marble. The room was originally furnished with inlaid tables and cabinets, stools covered with cut velvet or tapestry, and gilt-bronze candelabra. The original marble flooring and doors were replaced by the present wooden floor and doors in 1684.

14. **Franz Anton Bustelli.** *Columbine.* *c.* 1760. Nymphenburg porcelain. 7¾ in. high (19.7 cm.). Victoria and Albert Museum, London. This is one of a series of sixteen figures from the Italian *Commedia dell' Arte* designed by Franz Anton Bustelli, master-modeller at the Nymphenburg porcelain factory (until 1760 at Neudeck, a suburb of Munich) from 1754 until his death in 1764. Bustelli's new rococo style, all pointed elegance, un-inhibited emotion and wit, made Nymphenburg, under the patronage of the Elector of Bavaria, into one of the two or three leading porcelain works in Germany after the decline of Meissen. The vivacity of the pose, with its under-lying wistfulness of mood, is a perfect embodiment of the *Commedia dell' Arte* spirit—the subversive charm of im-poverished Italy brought to the formal, prosperous north, that charm which delighted artists and musicians from Watteau to Mozart.

15. (opposite, above). **Lukas von Hildebrandt.** *The Upper Belvedere*, Vienna. 1721–24. There is a hint of Versailles in this baroque summer palace, built for Prince Eugene of Savoy. (No great patron of the period, least of all one who had helped to defeat Louis XIV, could overlook such an example). But the architecture is more playful and festive. Hildebrandt is noted for his use of colour and variegated skylines. The arrangement of 'Upper' and 'Lower' Belvedere, with a long, stepped garden in between, echoes the 16th-century Belvedere in the Vatican.

16. (opposite, below). **Lord Burlington.** *Chiswick House*, London. *c.* 1725. This shows the English type of the Palladian summer villa, built by and for Lord Burlington, arbiter of taste in England in the second quarter of the 18th century: unostentatious, comparatively small, architecturally 'correct', picturesque only in its setting—in short, a reaction against everything that Versailles (and, nearer at home, Blenheim) stood for. The design recalls the villas built by Palladio in the Veneto in the 16th century, but modified in the light of Burlington's knowledge of Palladio's own reconstructions of ancient Roman Buildings.

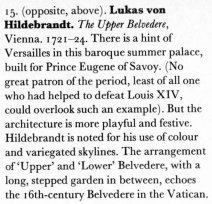

17. **Ignaz Günther.** *Virgin of the Annunciation* (detail). 1764. Carved and painted wood. 63 in. high (160 cm.). Weyarn, Upper Bavaria. The group of the Virgin and Angel of the Annunciation, of which this figure is a detail, was made for the Brotherhood of the Rose-garland at Weyarn to be carried in their processions. Seen as a whole, it is a composition of immense decorative sophistication, the turning poses of the figures interlocking with and answering each other, yet not touching, like the parts of a Bach fugue. The style is attentuated, refined, almost Gothic in its elongations, yet emotionally unrestrained and rococo. In feeling the analogy with Bustelli (plate 14) is close. Yet Günther, the supreme exponent of German rococo sculpture, lived in Munich the humble life of an ordinary master-craftsman.

18. **Christian Jorhan the Elder.** *St Nicholas* (detail). *c.* 1760–67. Carved and painted wood. Altenerding, Bavaria. 18th-century Bavarian sculpture offers a unique combination of sophisticated style and popular feeling. Jorhan is typical of the dozens of sculptors who produced work in this vein. He was influenced by Günther but never became so extreme.

19. (above). **Antoine Watteau.** *Les Plaisirs du Bal.* 1719. Oil on canvas. 19¾ × 24 in. (50 × 61 cm.). Dulwich College Picture Gallery, London; reproduced by permission of the Governors of Dulwich College. This record of the pleasures of a highly civilised country life was taken as a historical document by Hazlitt, who saw Louis XIV as the leading dancer, but, as with all Watteau's work, the setting and the vaguely 16th-century costumes put the picture into the realm of arcadian fantasy. For Watteau, as for Cocteau in 'Beauty and the Beast', the statuary is a living and participating element in the setting, and the landscape, taking up and tidying elements in Watteau's most influential forerunner, Rubens, never threatens to be more than an agreeable background to the fête.

in a picture of *St Martin dividing his Cloak*, although, unlike a baroque artist, El Greco still held the forward movement in check. It was taken up by Rubens in his portrait of the Duke of Lerma (1603), when it was given its first, experimental baroque form. Suaver, more accomplished interpretations of the motive occur in Rubens's later work and it received its final, fully baroque rending in equestrian portraits of Charles I by van Dyck. But illusionistic devices could also be used effectively in a much more restrained way, as in Rembrandt's *Lady with a Fan*. Here they are confined to the painting of the thumb of the right hand and the fan itself, which appear to overlap a false window-surround just inside the picture frame. Yet these touches are significant, however slight they may seem. Together with the 'close-up' placing of the figure and the intense gaze in the eyes, they remind us once more of the baroque emphasis on the quality of physical presence, an emphasis which is somehow made all the more telling in this instance by the very lack of baroque drama or movement. Rembrandt's sinister, watching figure seems almost bodily to have entered our world.

Of course, this impression is literally 'an illusion', and the question arises, how far is baroque illusionism intended to go? Illusionism is not *trompe l'œil*, although *trompe l'œil*—the actual tricking of the eye into assuming that a painted object is real—is no doubt illusionism in a specialised form. For practical reasons *trompe l'œil* is usually confined to still-life objects, whether treated as part of a picture or on a surface on which such objects might be expected to appear; the deception would hardly work with figures or landscape. Essentially, illusionism is intended to astound and to move, not to deceive. The expected response to it is similar to the way an audience responds to a play in the theatre. As members of the audience we know that the drama we are watching is not real, but we react to it emotionally almost as if it were real. From one point of view we remain conscious that what we are witnessing is art, yet in another sense we willingly suspend disbelief and allow ourselves to become involved in the situation. It is the

20. (left). **René Dubois.** *The Tilsit Table. c.* 1760–70. Oak, lacquered in green and gold *vernis Martin* and partly gilt, with chased and gilt-bronze mounts; the top inlaid with green Moroccan leather. Height, 30; width 56½; depth 28¾ in. (76 × 143.5 × 73 cm.). Reproduced by permission of the Trustees of the Wallace Collection, London. This table seems to have been made for Louis XV, who gave or sold it to the Empress Catherine II of Russia. It takes its name from the supposed fact that the Franco–Russian Treaty of Tilsit (1807) was signed at it, its then owner, Prince Kourakin, being the Russian Ambassador to the French Imperial Court. The intricate raised ornaments and elaborate caryatids on the legs (once attributed to the sculptor, Etienne Falconet,) give this piece a rich and rather sombre quality, for all its elegance, though the lacquer has darkened. The rectangular top and straight legs below the caryatids represent the beginnings of a classical reaction against the Rococo.

same with illusionism in the visual arts; indeed all representational art has this power to 'take us out of ourselves' in some degree, but illusionistic art is more consciously calculated to do it. St Teresa's mystical experience is vividly re-enacted in Bernini's statue before our eyes; Charles I in van Dyck's portrait rides out through a triumphal arch towards us (in its original setting in the now destroyed long gallery at St James's Palace this portrait probably filled the whole of one end wall down to the floor, thus intensifying the illusion which is now partly lost); in S. Ignazio we look upwards and see the roof of the church apparently open to the sky, its walls continued by an architectural construction painted in steep perspective, and hundreds of figures soaring upwards as far as the eye can see.

In each case we are moved to astonishment at the performance—a reaction that was certainly part of the artist's intention. The popular view of the Baroque is right in this

10. **Gianlorenzo Bernini.** *David slaying Goliath.* 1623. Marble. Life-size. Galleria Borghese, Rome. This is one of several marble statues made by the young Bernini for his chief patron in his early years, Cardinal Scipione Borghese, nephew of Pope Paul V. It illustrates the baroque principle of creating a focus beyond the limits of the work of art (the presence of Goliath is implied, not made actual in another sculpture across the room). The face is also a striking study in expression, for which Bernini is said to have used his own features studied in a mirror. The pose is adapted from the classical *Borghese Warrior*.

11. **Anthony van Dyck.** *Charles I with the Equerry, M. de St Anthoine.* 1633. Oil on canvas. 145 × 106 in. (368 × 269 cm.). Buckingham Palace (reproduced by Gracious Permission of Her Majesty the Queen). This equestrian portrait shows baroque illusionism and the mingling of the imagined and actual worlds in terms of painting. Charles I was one of the great collectors and patrons of the age, and van Dyck's portraits of him are among the 17th-century's most famous images.

respect: that surprise and spectacle play an important part in its total effect. And this part increased with time. By the early 18th century sheer spectacle had become the chief *raison d'être* of many works of art, for example Bavarian churches and palaces. Effects that had been evocative in the first half of the 17th century now became astounding. At the same time the element of make-believe in illusionism increased, although the actual use of illusionism in painting declined after 1700 except in ceiling decorations. Tiepolo's

8 ceilings are the supreme example. They are illusionistic on account of the fact that flying figures seen from below are represented in the sky but there is no longer any pretence that they are actually there. The whole thing has become an immensely sophisticated and exquisite conceit, as formal and self-conscious as Mozartian opera—and as dazzlingly beautiful.

THE FUSION OF THE ARTS

Just as the distinction tends to get blurred in baroque art between art and life, so there is often a blurring of the distinctions between the various arts. The arts were also apt to exchange roles, or rather each moved in the direction

of painting. Architecture became more sculptural, sculpture more pictorial and painting itself more strictly concerned with visual appearances; that is, by a greater stress on light, shade and colour rather than form and outline, painting was made to reproduce more faithfully what the eye sees as distinct from what it knows to be there.

Underlying all this and connected with the devices of illusionism and close-up discussed so far, was the tendency in baroque art to deny the importance of the frame. Compositionally, the interest was concentrated in the central area, thus detracting attention from the edges. In this connection the fullest advantage was taken of the way altarpieces were lit by candles from below so that highlights and prominent relief were emphasised while the shadows merged into the surrounding architecture.

In tombs, altars and decorative *ensembles* the three arts of painting, sculpture and architecture were increasingly combined. It is true that there are Renaissance precedents for this (e.g. the Chigi Chapel by Raphael in Sta Maria del Popolo in Rome), but such precedents are rare, whereas in the high and late baroque periods (less so in the early baroque) the amalgamation of the arts became the rule rather than the exception whenever it was appropriate.

In this respect, as in so many others, Bernini was the most original and creative mind. He had been a painter and stage designer when young as well as a sculptor and architect, and his experience in all four arts was fused in his imagination. Admittedly his early works scarcely show much inter-connection between the arts. His first sculptures are pure sculpture, his one early building (the church of Sta Bibiana) is pure architecture, and he did not begin to introduce pictorial effects into architecture or sculpture until he had given up pure painting. But in his mature works the fusion of the arts is complete. Sometimes sculpture, sometimes architecture predominates, but his build-

ings always include sculptural features (S. Andrea al **39** Quirinale) and his sculptures often have an architectural setting or base (the Tomb of Alexander VII.). Occasional- **85** ly, as in the *Baldacchino* in St Peter's (conceived by Bernini **12** though perhaps designed in detail by Borromini), it is impossible to decide to which category the work belongs. In every case the two arts are fully integrated, so that each is, as it were, a continuation of the other.

The contributions of painting and stage design to the result are found partly in the introduction of pictorial and theatrical elements, so that the work has something of the 'look' of a picture or a spectacle on the stage, and partly in the use of coloured materials: dark-brown and gilt bronze, and plum-red, brown, green and yellow-ochre in addition to white marble.

The two works which show these characteristics most clearly are the *St Teresa* and the *Cathedra Petri* in St Peter's. **13,1** The first is framed like a picture with marble columns at each side and a curved pediment above, while sculptured figures in niches let into the chapel walls sit and discuss the miraculous event portrayed on the altar, like spectators in

12. **Gianlorenzo Bernini and Francesco Borromini.** *The Baldacchino in St Peter's, Rome.* 1624–33. Bronze, partly gilt. 93 ft. high (28.5 m.). On account of its curving shapes, twisted columns and ornate splendour, this majestic canopy over the tomb of St Peter has been called the first manifesto of Roman high baroque art. It is an architectural work treated in sculptural terms; note the illusionism in the way the bronze is used for the cloth hangings at cornice level. The Baldacchino stands directly beneath the dome of the Basilica. The *Cathedra Petri* (fig. 13) can be seen in the background on the left. The twisted columns echo the late classical columns supposed to have come from the Temple of Jerusalem, which were built into the Old St Peter's and of which two can be seen flanking the niche at the top right.

13. **Gianlorenzo Bernini.** *The Cathedra Petri.* 1657–66. Marble, bronze (partly gilt), coloured glass and gilt stucco. St Peter's, Rome. The *Cathedra Petri* is so-called because its centre-piece encloses the ancient wooden chair believed to have been the throne of St Peter. The four great bronze figures supporting the chair are the Greek and Latin Fathers of the Church: from left to right, SS. Ambrose, Athanasius, Chrysostom and Augustine. The back of the chair is decorated with a relief of the scene 'Feed My Sheep'; the papal keys and tiara and, above them, the Dove of the Holy Ghost, in the centre of the window, are further important features of the scheme.

theatre-boxes. The validity of this comparison has recently been disputed, on the ground that the altar frame projects so far into space that the figures cannot see the saint; this is true but is not apparent to the real observer standing in the chapel. It has also been objected that most of the sculptured figures are not watching the 'stage', but this would probably have been typical of 17th-century theatre audiences.) The second work, the *Cathedra Petri*, has the whole west end of St Peter's for its stage (unlike most cathedrals and churches, St Peter's is orientated towards the west) and has no frame, but spills out irregularly over the two great pilasters, which belong to Michelangelo's architecture, at either side of the round window in the centre.

Nor is this all, for Bernini did not only make use of more than one art for these works; he also harnessed nature to his ends. In both cases light, filtered through yellow glass and continued illusionistically by means of gilt metal rays, is made part of the work of art. In the *St Teresa* it comes from a concealed source above the altar; in the *Cathedra Petri* it streams through the central window, which contains an image of the Holy Dove in leaded outline and is the focal point of the design. Surrounding this window are figures of

15. **Jean Baptiste Tuby** (after a design by **Lebrun**). *The Fountain of Apollo.* 1668–70. Lead. Engraving by Chastillon, 1683. Versailles Gardens. This fountain is the centre-piece of a pool linking the end of the *tapis vert* with the head of the *grand canal* and is one of the most spectacular pieces of garden sculpture at Versailles. The surfaces of the figures and horses were originally covered with gilt bronze. Fountains were an integral part of the conception of Versailles from the start and much ingenuity and money were expended on organising the water supply.

14. **Gianlorenzo Bernini.** *The Four Rivers Fountain.* 1648–51. Stone (travertine), the figures in marble. Engraving by G.B. Falda. Piazza Navona, Rome. The obelisk is a restored classical remain which was found in the Campagna and which Pope Innocent X ordered to be erected in the centre of the square which contained his family palace. The fountain was designed by Bernini, although the base and figures were executed by assistants. The four personified rivers are the Nile, Danube, Ganges and Plate. The church of Sta Agnese (pl. 84) is just outside the picture to the left. The Piazza Navona is the largest and most famous baroque square in Rome.

angels in gilt stucco floating on clouds. Below them is suspended the magnificent bronze throne containing the relic—the ancient wooden chair of St Peter—from which the *Cathedra* gets its name. On the lowest level, supporting the throne with miraculous ease, are four colossal bronze figures representing the Fathers of the Church. The effect created by this astounding work, seen framed between the giant twisted columns of the Baldacchino in the late afternoon when the setting sun comes into line behind the window, is one of the supreme visual triumphs of the Baroque.

As an alternative to natural light as part of a work of art there was also water. It goes without saying that the fountain as a form of decorative art was not a baroque invention, but fountains now became larger, freer in design and more illusionistic in treatment. The conceit inherent in the idea of a fountain—that of a fictitious figure immersed in a real element and performing the real act of spouting water—was exploited in ever more ingenious ways. Mythological and allegorical figures, sea-horses, cherubs, dolphins, etc., associated with water, lent themselves naturally to the purpose. In one instance, Bernini's *Neptune*, which actually stood above, rather than formed part of, a fountain, the figure is jabbing at the water with

his trident, illustrating a passage from Virgil's *Aeneid* describing Neptune calming the waves.

But this narrative treatment, though characteristic of Bernini's artistic personality, was unusual. More often the fountain was a purely decorative feature, the symbolism of which was not intended to be interpreted too precisely or taken very seriously. The best known fountains used in this way are those in the gardens at Versailles, where their function is to bridge the gap between the contrived, man-made splendour of the château and the informal beauty of nature. Although in one sense the fountain is a minor art form, from another point of view it can be regarded as the *ne plus ultra* of the Baroque, on account of its fusion of the two arts of sculpture and architecture with each other and of nature with both. Of Bernini's *Four Rivers* fountain in the Piazza Navona one may fairly ask—is it sculpture? is it architecture? is it nature?—and get no clear answer.

Following Bernini's example, the use of two or more arts in combination, with or without the addition of natural light, was taken up by countless baroque artists in Italy, Austria, Southern Germany and Spain (less so in Northern Europe and France). Gaulli's illusionistic ceiling of the Gesù, the composition and outline of which directly echo the *Cathedra Petri*, has a feigned not a natural source of light and most of its figures are painted, but the painted surface spreads across the frame on to the coffered vault of the church and some of the figures towards the edge are treated partly in painting and partly in high-relief stucco. From the ground one can hardly tell where the painting ends and the real stucco and architecture begin.

As to the later use of natural light, the best known example is probably the *Transparente* in Toledo Cathedral by Narciso Tomé, which is even more spectacular than the *Cathedra* though less subtle in design. Here the source of light is not within the *Transparente* itself (and is not visible in the part of the work shown in plate 2) but above and facing it across the open space of the Gothic ambulatory of the cathedral. The light, that is to say, comes from a window let into the vault opposite the altar and is only seen if the spectator turns and looks over his shoulder. But the vault itself is filled with sculptured figures of prophets and saints in the same style as those on the *Transparente*; they thus form part of the same work of art. The asymmetrical relationship of the two parts, with one on a different level from the other, together with the upward three-dimensional movement of the composition, are typical of the late Baroque.

Also typical of the late baroque phase is the way in which pure spectacle in the use of these devices tends to outweigh all other considerations. This is especially true of German baroque art. In Egid Quirin Asam's *St George* altar at Weltenburg the statue of the horse and rider are seen head-on against a picture painted on the back wall, which is some distance behind the altar itself. The space between is flooded with light, which throws the statue into startling silhouette and gives it a flickering, flame-like outline while leaving its internal forms mysterious and dark. The result is frankly theatrical, much more so than was the case with Bernini's *St Teresa* or *Cathedra Petri*, in which methods borrowed from the theatre were used not as ends in themselves but in order to put across a serious message with the maximum effect. It is important to realise that there are degrees of theatricality in baroque art. Bernini's ingenious but controlled use of stage effects belongs at one extreme; at the other there is the mid-18th-century Frauenkirche at Dresden, which is designed like a theatre throughout, with the seats arranged in a semi-circle round the altar space, and galleries all round, as they would be in an auditorium.

By this time theatre design had itself become a highly sophisticated art in its own right. From Bologna a whole family of artists, the Bibiena, gained international reputations by their stage designs alone, and the architect, Juvarra, who produced his greatest buildings in north Italy, also worked in this field. More than one factor drew the theatre and baroque art together. There was the love of display which was common to both; there were the opportunities for controlled lighting effects, illusionism and trick perspective; the theatre, like the pulpit, was the natural home of rhetoric. Finally, the theatre was a 'total art' in a way that visual art could never be, as it involved real figures

16. **Georg Bähr.** *The Interior of the Frauenkirche, Dresden* (destroyed in the Second World War). Begun 1726. Although built and always used as a Protestant church, the Frauenkirche in Dresden was structurally as baroque as the Catholic churches of Bavaria, particularly in its analogies with the theatre. The spaces leading off the central space and seen through the arches can be compared to the stage design by Bibiena, fig. 17. The very beautiful organ case at the back of the 'stage' stands in for the altar. Both the placing of this organ and the multiplication of subordinate spaces and singing galleries reflect the growing importance of church music in Germany in the age of Bach.

17. **Giuseppe Galli Bibiena.** *Design for a Stage Set*. Engraving, by J. A. Pfeffel, from the series, *Architettura e Prospettive*, Augsburg, 1740. 12 × 20 in. (30.5 × 50.7 cm.). Stage design was in a sense the logical end to which baroque art tended, since its virtual freedom from material and structural limitations made possible the realisation of effects, particularly of illusionism and perspective, that Bernini or Borromini could never have created in marble or stone. Although based on Bologna, the various members of the Bibiena family worked as much for Austrian and German as for Italian courts, designing settings for operas and shows connected with betrothals, weddings and other celebrations.

18. **Annibale Carracci.** *The Farnese Ceiling* (detail of corner).
1597–*c.*1600. Fresco. Palazzo Farnese, Rome. The Farnese
Ceiling, painted on the flattened barrel vault of the main, 66 ft
long, gallery of the palace for its then owner, Cardinal Odoardo
Farnese, was the first great painted ceiling of the baroque
period. Its subject was the loves of the Gods in Antiquity–an
unusual choice for a Prince of the Church, although such a
subject was just then becoming possible in the more
enlightened atmosphere that set in around 1600. The pictorial
arrangement is based on a painted architectural framework open
to the sky at the corners; on this framework are painted imitation
stone figures (herms), imitation bronze medallions, imitation
pictures (at left of this detail) and 'real' nude figures kneeling or
seated on the actual cornice of the room. The ceiling is thus
illusionistic in detail, if not as a whole, as well as being highly
decorative.

19. **Annibale Carracci.** *Study for One of
the Nudes in the Farnese Ceiling.* 1597–98.
Black chalk with white heightening.
16¼ × 16⅛ in. (41.3 × 41 cm.). Louvre,
Paris. Annibale Caracci is thought to have
made over 600 drawings in preparation
for the Farnese Ceiling, ranging from
rough sketches of whole compositions to
handsome studies for individual figures, of
which this is an example. In thus preparing
the work in great detail he revived the
practice of Raphael and Michelangelo,
which had been in abeyance during the
Mannerist period, and handed this
practice on to his successors.

(actors) and speech in addition to costumes and scenery.
Extending its range still further, the theatre branched out
during this period into opera—potentially the 'total artwork' *par excellence.* The first opera was actually performed
in the late 16th century—a fact which is sometimes over-
looked by those who claim that the invention of opera was
'a product of the Baroque'. Nevertheless, opera was an art
which only spread from its original home, Italy, to the rest
of Europe in the 17th century and only became fully developed during that period. It could fairly be said that
opera and baroque art evolved at more or less the same
time and speed and, in some important respects, in similar
ways.

THE SPLENDOUR OF BAROQUE ART

No account of the Baroque would be worthwhile that did
not include some discussion of its qualities of sheer visual
splendour. Baroque artists appealed to the spectator not
only through the use of illusionistic devices, methods borrowed from the theatre and eloquent gestures and expressions, but also by providing enchantment for the eye. Except for the outsides of buildings, which were in natural
stone, marble or brick, they used rich colours in all the

arts: coloured marbles and coloured or gilt carved wood in
interiors, and saturated colours in paintings. Rooms in
palaces were filled with expensive furnishings and tapestries; even church interiors became increasingly ornate.

Some of this visual magnificence was achieved merely by
accumulating the maximum quantity of material. This
was especially the case with Spanish architecture and decorative art (though not with painting), whereas the
design itself is often uninventive. But in Italy, France and
Germany, the ornamental quality of art is not something
added afterwards but an integral part of the conception.
The design, whether in the style of the figures in painting
and sculpture, or in the treatment of columns, capitals,
windows, etc., in architecture, has an element of sensuous
beauty which attracts in itself. This is perhaps most obvious
in Austrian and German churches and palaces of the 18th
century, but it can also be seen in a restrained way in the
first masterpiece of early baroque ceiling decoration,
Annibale Carracci's ceiling in the Palazzo Farnese, begun *18*
in 1597.

If one compares this ceiling with corresponding Renaissance works, from which it was ultimately derived, the
main difference lies precisely in its bias towards decoration.

The austerity of Michelangelo's *Sistine Ceiling* is replaced by a new lightness of conception and touch and a greater use of decorative assessories. Painted shells, masks, swags of fruit and other ornaments cover the feigned architectural framework. Just how deeply this emphasis on sensuous visual beauty goes can be seen by comparing Annibale Carracci's drawings with Michelangelo's. The latter are hard, austere and sculptural. The former show a conception of the body that is altogether more sensuous; the figures are more fluid in outline and seem literally to breathe a sense of physical well-being.

The same qualities lie at the heart of Rubens's style both as a draughtsman and painter. His feeling for sensuous beauty is not only superficial; it does not lie only in the saturated colours he uses, the opalescent brilliance of his flesh, or the luminous shadows and expensive dress materials which he liked painting. It is an essential quality of his whole art—a sort of energising centre nourished by the physical powers of the body—of which these outward manifestations are only the sign. He transmitted the outward manifestation, though less the vital centre, to his successor van Dyck. In van Dyck's portraits the self-conscious cultivation of luxury and the arts of personal adornment assumes the status of a way of life.

One form of the love of splendour in architecture can be seen in the treatment and placing of buildings in such a way that they gain the best advantage from a superlative site; examples are the monastery of Melk overlooking the Danube, and the church of Sta Maria della Salute, Venice, which guards the entrance to the Grand Canal and 'composes' the view from whichever direction one looks at it. The conscious creation of a style to exploit these natural advantages was a special feature of north Italian, Austrian and south German architects of the late baroque period (although Sta Maria della Salute is mid-17th-century).

Another manifestation of the same attitude was the taste for magnificence on a very grand scale, independently of the site. By no means all the greatest or even the most dazzling works of baroque art are large; there is a jewel-like richness even in quite small buildings like Borromini's church of S. Carlo alle Quattro Fontane and also, of course, in works of the applied arts; jugs, table ornaments, candle-sticks, doorknobs, etc. Nevertheless, grandeur for its own sake is obviously a characteristic to which the Baroque tends. St Peter's, Rome, in its present, baroque form is one of the largest buildings in the world, although it should be remembered that Michelangelo's contribution to it was already on a colossal scale. In painting, Rubens prided himself on his ability to undertake vast decorative projects; the forms in his larger pictures are the embodiments of majesty and power; and his handling in these pictures has an incomparable breadth and sweep.

In secular art the taste for grandeur was given a new impetus by the creation of Versailles, where everything was done in the most spectacular way possible. In its finished form, for which Jules-Hardouin Mansart was re-

20. **Pierre Patel the Elder.** *View of the Château of Versailles* (detail). 1668. Oil on canvas. $45\frac{1}{4} \times 63\frac{1}{2}$ in. (115×161 cm.). Musée de Versailles. Begun in 1624 by the architect, Le Roy, as a small country residence for Louis XIII, Versailles was transformed, first through the enlargements begun by Le Vau in 1669 and secondly through the still vaster additions undertaken by J.-H. Mansart from 1678 onwards, into the grandest palace in Europe. Patel's painting shows the Château as it was completed for Louis XIII, although the ranges either side of the forecourt (now the *Cour Royal*) may have been added by Le Vau when he first started working there for Louis XIV in the early 1660s.

21. *Plan of the Château of Versailles*

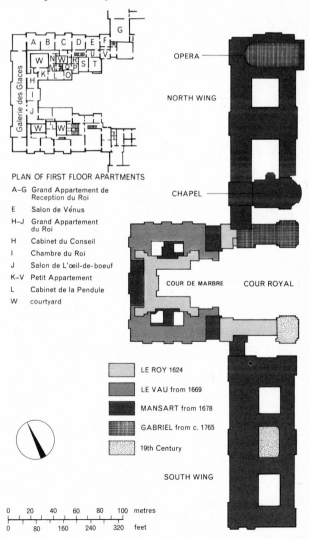

PLAN OF FIRST FLOOR APARTMENTS

A–G   Grand Appartement de Reception du Roi
E   Salon de Vénus
H–J   Grand Appartement du Roi
H   Cabinet du Conseil
I   Chambre du Roi
J   Salon de L'œil-de-boeuf
K–V   Petit Appartement
L   Cabinet de la Pendule
W   courtyard

OPERA
NORTH WING
CHAPEL
COUR DE MARBRE   COUR ROYAL

LE ROY 1624
LE VAU from 1669
MANSART from 1678
GABRIEL from c. 1765
19th Century

SOUTH WING

0   20   40   60   80   100 metres
0   80   160   240   320 feet

12 sponsible, the Château is nearly 600 yards long. Inside and outside, teams of craftsmen were employed under the overall supervision of Lebrun to make the Château the most magnificent palace in Europe, which it undoubtedly remains. Not that the architecture as such is particularly baroque, apart from the chapel; if anything, it is classical in form and disposition. But its cumulative effect—a conscious exercise in all the techniques of display—is certainly baroque in feeling. This is partly a result of the scale and massiveness of the architecture, but it also derives from the

11 baroque treatment of the accessories and furnishings. This contrast between outside and inside, and between the major arts of painting, sculpture and architecture on the one hand, and the minor arts of furniture, silverware, textiles, etc., on the other, is often seen in classical contexts.

Returning to painting, a word should now be said about brushwork. For a variety of reasons it is no accident that the pictorial approach of baroque painters was derived more from Venetian than Florentine or Roman Renaissance art. In the 16th century it was above all in Venetian painting that the qualities of sensuous beauty and splendour associated with the Baroque were found. A crucial moment in the history of the birth of Baroque occurred in 1598 with the transference from Ferrara to Rome of a group of mythological paintings by Titian. These paintings were studied by almost every artist working in Rome from the time of their arrival there until the 1630s; among those who copied them were Rubens and Poussin.

In the 17th century itself, Venetian painters contributed little to the Baroque, but the influence of 16th-century Venetian art on baroque painting, not only in Rome but elsewhere in Italy, and still more in Flanders and Spain (Rubens had studied in Venice and Rome, Velasquez had seen many Venetian paintings in Madrid), was immense. Without doubt the most important forerunner of the Baroque in painting was Titian. Titian's art had the power and the mastery of colour, light and shade that baroque painters adopted.

The free, expressive brushwork of Venetian painters was no less influential. A broad, lively and beautiful handling of paint is one of the most widespread characteristics of 17th-century art. There were no rules for this and each artist developed his own individual style. For instance,

3 there was the fluent, dynamic handling of Rubens, in which the brushstrokes seem to record every movement of the

23 artist's powerful mind. There was Velasquez's dryer, less personal stroke which gives an impressionistic texture to the surfaces of his canvases and an independent beauty to colour which is unique in the 17th century. There was

25 Frans Hals's method, which is superficially similar to Velasquez's, but more vigorous and cruder.

22 Finally there was Rembrandt, whose brushwork is the most expressive and varied of all. Rembrandt's handling is sometimes far from attractive in the ordinary sense. Sometimes it is harsh and expressionistic; sometimes its beauty lies in its representational subtlety; but sometimes it has a

22. **Rembrandt van Rijn.** *Portrait of Jan Six* (detail). 1654. Oil on canvas. 44 × 40 in. (112 × 102 cm.). Jan Six Collection, Amsterdam. This example of Rembrandt's brushwork dates from the later part of his career, when he had ceased to model forms by tracing their outlines and had turned instead to using irregular patches and smudges of paint to fix the positions of forms and to indicate the direction of their lines of force. No brushwork in European art is more subtle and expressive than this.

23. **Diego Velasquez.** *Portrait of a Lady with a Fan* (detail). *c.*1635. Oil on canvas. 36½ × 27 in. (92.8 × 68.5 cm.). Wallace Collection, London. Velasquez's brushwork is more elegant and regular than Rembrandt's, but both artists used basically the same principle—that of expressing colour, form, atmosphere and tone by means of the brushstrokes alone. This means that the brushwork is not just something added to a previously drawn and shaded design but is an integral part of the conception.

richness of texture that is as glowing and sumptuous as that of any 17th-century painter. These four artists are no doubt conspicuous, but the feeling for paint as a medium and the optical approach to form, stressing colour, light and shade rather than outline, affected all artists of the period in varying degrees. Even though classical theory condemned this characteristic and criticised the appeal to the senses as morally wrong, those artists who were most sympathetic to classicism unconsciously found themselves using richer colours and freer brush-strokes than their Renaissance predecessors.

BAROQUE DRAMA AND THE USE OF LIGHT AND SHADE

On the one hand, drama in baroque art is connected with the psychological and theatrical devices already described; on the other, it is a function of the effects of movement to be discussed shortly. In fact like most characteristics of baroque art, its qualities of drama merge into those on either side of it and would hardly need to be treated separately except for their associations with another quality —violence—and the treatment of light and shade. Most baroque paintings have dark backgrounds; their colours are rich and glowing rather than bright; the light in them tends to be concentrated on forms in or around the centre of the composition. The same applies to architecture, in that churches especially tend to be unevenly lit. Windows are high and often small, and most of the daylight comes through a lantern in the dome. The result is an evocative, sometimes dramatic chiaroscuro effect similar to that in paintings. The light may be quite strong in some parts but deep pools of shadow are left in the recesses, in doorways, under arches, behind pillars, etc.. To this fitful natural light and shadow must be added the evocative light of candles. It was only in the 18th century that churches became light and airy once more.

40 The first painter to use strong contrasts of light and shade, and to introduce a special emphasis on violence in art, was Caravaggio. He made the two mutually supporting. In his work the large areas of almost black shadow not only close off the background, but heighten the drama inherent in the subject-matter; even where there is no actual violence the darkness creates a mood of foreboding and tragedy. It is a mood similar to that evoked by Webster's lines in the *Duchess of Malfi:* 'Oh this gloomy world, / In what a shadow, or deepe pit of darknesse / Doth (womanish, and fearefull) mankind live!'

Caravaggio's direct influence was brief, though intense, and was confined to his immediate followers, many of them foreign-born, who worked in Rome. But the indirect consequences of his work for European art were far reaching and incalculable; Velasquez in Madrid and Georges de la Tour in Lorraine probably produced the finest and most sensitive of all paintings in the Caravaggesque style, although neither of them may ever have seen an original work by Caravaggio. In a wider sense Caravaggio's innova-

tions spread even further and percolated into still-life and *genre* painting.

On the other hand, the pervasive current of violence and darkness in 17th-century painting was certainly not due to him alone; there were other methods of creating strong contrasts of light and shade. Like brushwork and colour, those methods were derived from 16th-century Venetian painters—to some extent from Titian but still more perhaps from Tintoretto. The Venetian treatment of light and shade, which was more luminous and atmospheric than Caravaggio's, is clearly seen in the work of Bolognese painters like Guercino. It also occurs, applied in a more **46** decorative sense, in the paintings of Rubens and van **3,11** Dyck. Neopolitan painters used a Caravaggesque type of chiaroscuro softened slightly by free baroque brushwork and varied by half-lights (Ribera, Cavallino). *24*

However, the greatest master of chiaroscuro was, once *70,***7,26** again, Rembrandt. He may have owed something in- **27,48** directly to Caravaggio and something to the Venetians. Although the never left Holland he made himself aware— partly by attending auction sales, partly by studying the work of his lesser but more cosmopolitan Dutch contemporaries—of the whole heritage of European painting since the Renaissance. But the final result was his own. Like Caravaggio he tended to concentrate the light on the forms in the centre of the composition and to leave the background in darkness. But whereas Caravaggio's darkness is space-denying, Rembrandt's is space-creating. Varied by half-lights and reflected lights and also sometimes by light from more than one source, it is intensely luminous. Its effect is spiritual and poetic rather than dramatic.

But Rembrandt, too, sometimes painted scenes of violence, especially in the relatively early period of his career following his move to Amsterdam in 1631. Occasionally there is more than a hint of sadism in these works. The *Night Watch*, too, is a painting of great dramatic power, **7** though the drama lies in the method of presentation rather than the subject: light glows in the centre of the scene, drums beat, banners are unfurled and the members of Captain Frans Banning Cocq's militia company march out on parade in a swirl of movement. Drama and splendour are here combined.

To Catholic painters, scenes of martyrdom gave further *24,25* opportunities for the representation of violence. Such scenes were nearly always shown in semi-darkness, which intensified the horror—in fact it was often the chief means of expressing it. The overt dwelling on the physical effects of torture—bloated bodies, torn limbs and streaming blood, which are common in German Renaissance art—is less a feature of baroque martyrdoms than one might expect. This was because the paintings had a religious purpose corresponding to the doctrines of the Counter-Reformation. The martyrs were represented as heroes, not as the limp, broken, helpless victims they would have been in fact. Physical agony was softened by spiritual exaltation in their expressions.

24. **Jusepe de Ribera.** *St Sebastian tended by St Irene.* 1628. Oil on canvas. 61½ × 74 in. (156 × 188 cm.). Hermitage, Leningrad. The surface realism and strong tone contrasts characteristic of Caravaggio are here combined with baroque plasticity and breadth of handling. This 'baroque' interpretation of Caravaggio may be compared with the classical interpretation seen in Georges de la Tour's rendering of the same subject (fig. 51). A certain voluptuousness was traditional in the treatment of St Sebastian's martyrdom but the dwelling on the spiritual exaltation rather than the physical agony was typical of the 17th century. In fact, the episode in which the saint was nursed back to life by St Irene and her attendants was more often painted in this period than the scene of his 'execution' by archers.

25. **Salvator Rosa.** *Landscape with the Crucifixion of Polycrates. c.* 1665. Oil on canvas. 29 × 39 in. (73.5 × 99 cm.). Art Institute of Chicago. A taste for the macabre, which was endemic in Neapolitan painting, already appears in Ribera's work and was brought to an extreme by Salvator Rosa. This can be seen both in his figures and his landscapes, which, after Rubens's, are the most baroque in 17th-century painting. The strong tone contrasts, jagged rocks and shattered trees are typical. Polycrates was the ruler of the island of Samos and a patron of poets and artists. He was lured to the mainland by Oroetes, the lord of Sardis, and there crucified in 522 BC. The subject is very rare in art.

It would be wrong to condemn this approach as insincere or self-indulgent, as if it were merely a dishonest attempt to make death by martyrdom seem attractive. Apart from the religious purposes involved, the painters were concerned with exploring states of mind. Nor was the agony of the martyr the only state of mind to be represented. In baroque religious art we find ecstasy, adoration, sorrow, triumph, wonder, contemplation and many other moods, as well as hatred and pain. One of the most eloquent representations of emotion in the art of the period is also the subtlest and most restrained—Rembrandt's *Bathsheba*.

48

BAROQUE MOVEMENT

The characteristics typical of the baroque treatment of movement, which are as important for architecture and sculpture as they are for painting, hardly appeared before the 1620s. The chief problem for baroque artists before that date was to substitute coherence and correct proportions for the anarchy prevailing under Mannerism. The burden of solving this problem was mainly carried by Italian artists of the generation of Caravaggio, Annibale Carracci and Maderna. In most other countries late 16th-century styles persisted until the second quarter of the 17th century.

The first painter to develop the techniques of true baroque movement was Rubens, who had lived in Italy from 1600–1608 and was thoroughly familiar with the whole of 16th-century and contemporary Italian art. His great discovery, which he made a few years after returning to Antwerp in December, 1608, was the principle of building his compositions round a dynamic spiral line. This line is a line of force which attracts all forms to it. It is essentially three-dimensional, starting at the bottom in the foreground of the composition, either at the centre or to one side and soaring upwards and inwards in a spiral or zig-zagging diagonal. Its great and novel consequence is that it gives an unprecedented vitality to the whole picture, which is now conceived in dynamic not static terms. Moreover, it is purposeful movement which leads up to a climax; it is quite different from the scattered hither-and-thither move-

26, 2

ment of Mannerist art, which has great vitality but leads nowhere.

Nor does Rubens's three-dimensional line apply only to the composition; it has its counterpart in every figure and every detail of the picture. Figures surge, twist and turn, and draperies flow, not in an altogether natural way, but not in a contorted way either, so that once the eye accepts the artificial principle on which the painting is conceived —and it does this quite easily—the movement appears logical and convincing. It is a movement which is reflected not only in the poses of figures but also in their modelling and the contours of their limbs. It is as apparent in oil sketches and drawings as it is in finished pictures; in fact, it is seen at its purest in the sketches, where a figure in bold foreshortening and of the utmost complexity may be clearly indicated with no more than half-a-dozen fluent strokes with a fine brush.

Rubens was also the creator of another characteristic of baroque movement: the tendency for forms to blend and merge into one another. In a full baroque painting the eye is given only a few points of rest. Apart from these points the forms are not distinct from one another as they are to a large extent in Renaissance art; rather, each one leads into the next. The relationship of forms is like the sequential arrangement of images in Milton's poetry. One can hardly dwell on each in isolation, but must take them as part of a continually moving whole.

It was not until the 1620s that Italian artists reached an equivalent stage. The innovator here was Bernini in a series of sculptures mostly done for Cardinal Scipione Borghese (the *Apollo and Daphne*, the *David* and the *Neptune*). As one might expect from an Italian artist and a sculptor, the forms are more solid and compact than is the case with Rubens's paintings, but the principle of movement is similar. There is the same thread of energy running right through the figures and the same power of conceiving it in three dimensions. The result is once more essentially dynamic, not static, and the forms flow organically into one another. It is the same in later, more complex works like the *Cathedra Petri*, in which the various parts of the composition merge into one another. The climax of Bernini's method can be seen in one of his last works, the *Angel with the Superscription*. From a formal point of view there is scarcely any distinction between the treatment of the drapery and that of the figure. The whole sculpture is animated by a refined, sophisticated, almost feverish vitality, which runs down into small details like the chiselling of the hair and the fingertips.

It was from such works as this that Austrian and German 18th-century sculpture was evolved. But by then the movements had become quicker, suaver and more graceful; less energy was released as the forms became lighter in weight. Bodies were made thinner, and heads, now smaller than before, were turned at a sharper angle to the shoulders—in paintings they were often turned away from the spectator and their faces were lost in shadow. Figures seem all

26. **Peter Paul Rubens.** *The Conversion of St Paul. c.* 1616. Oil on panel. $37\frac{1}{2} \times 47\frac{1}{2}$ in. (95.2×120.7 cm.). Count Antoine Seilern, London. Starting from the point where Leonardo da Vinci and Titian had left off, Rubens transformed the traditional pictorial theme of struggling horses and riders into the earliest manifestation of high baroque painting in Europe. The composition is based on a series of contrasting curves held together by a powerful 'S' which spirals upwards and inwards from the bottom left. Note also the violent foreshortenings typical of the Baroque.

27. **Peter Paul Rubens.** *Landscape with the Shipwreck of Aeneas.* 1624. Oil on canvas. $23\frac{1}{2} \times 38\frac{1}{2}$ in. (60×98 cm.). National Museum, Berlin-Dahlem. Rubens was the greatest master of baroque landscape in the first half of the 17th century. Natural forms—trees, rocks, streams, paths and so on— are subjected to the same twisting dynamic movements and contrasting light and shade patterns as are human beings in his figure compositions.

28. **Egid Quirin Asam.** *The Assumption of the Virgin* (detail). 1718–22. Painted stucco. Life-size figures. Monastery Church, Rohr, Bavaria. Baroque principles of modelling underlie this group by Asam, but the tortured energy of Bernini or Puget has been replaced by an air of theatrical ease; the Virgin gestures and opens her mouth like an opera singer and is seen to have attained her position not by her own efforts but with the help of some supernatural agency. The outline of the group no longer follows the contours of the human body but is defined by limbs, wings and fluttering draperies. From this 'late baroque' point it was a short step to the Rococo.

29. **Giovanni Battista Gaulli.** *The Adoration of the Name of Jesus.* 1674–79. Fresco. Ceiling of the Nave of the Gesù, Rome. This fresco represents the transition from the 'high' to the 'late' Baroque in Roman ceiling decoration. Compared with fig. 30, the whole ceiling is treated as a single unit of space, without any architectural framework, and the painted figures are made to spread out onto the coffered vault of the church itself; the illusion is that of a roof opened to the sky to reveal a vision of the heavenly host in adoration of the Holy Name, while damned souls tumble into the darkness below. Both the figure style and the irregular outlines of the composition echo the *Cathedra Petri* (fig. 13). See fig. 2 for a view of the interior of the church.

limbs. Legs and arms often trace lines so complicated that the spectator can hardly work out how the figures are posed. Movement in the spaces between forms was also indicated with a new subtlety. The Virgin and the Angel in **18** Günther's sculptured group of the *Annunciation* are as intricately related to each other by their movements as two figures in a dance, although they never touch. The wider gaps left between figures in late baroque ceilings also allowed new scope for implied movement across open spaces. **8** Tiepolo, seizing on the possibilities of this, exploited it with consummate ease and grace.

Movement in Italian mid-17th-century baroque ceiling paintings seems heavy by comparison with Tiepolo's soaring flight; it is like an irresistible force whose characteristic is power rather than speed. It also stops short of the treatment of details, which are handled almost as if they were static forms. In Guercino's *Aurora*, where baroque movement is fully manifest in Italian painting for the first time, **46** Dawn in her chariot charges across the sky, scattering flowers and expelling the fleeing figure of Night. It is not so much the horses, which appear to have stopped in mid-flight, as the subsidiary forms—the clouds, the flowers and the figure of Night—which bring the composition to life.

A more dynamic treatment of form can be seen in the next great Roman ceiling painting, Pietro da Cortona's *Triumph of Divine Providence* in the Palazzo Barberini. Here **30** the figures surge across the surface in a vast, almost impenetrable mass moving in different directions and sometimes getting in each other's way, so densely are they

30. **Pietro da Cortona.** *The Triumph of Divine Providence.*
1633–39. Fresco. Palazzo Barberini, Rome. This vast and
complex fresco, painted on the ceiling of the *Gran Salone* in the
family palace of the reigning pope, Urban VIII, inaugurated a
new phase in Roman baroque ceiling decoration. The painted
architectural framework and many of the decorative details at
the corners were adapted from the Farnese Ceiling (fig. 18), but
a new dynamic surges through the composition, the spaces are
more open and all the forms are shown in writhing movement
which focuses on a single figure, the personification of Divine
Providence. In detail, the meaning of the work is highly
complicated but its central theme, implied rather than stated, is
the presentation of Urban VIII as the agent of Providence.
The fresco is thus a specimen of propaganda on behalf of 'divine
right' comparable to the almost contemporary Whitehall Ceiling.

31. **Pierre Puget.** *Herm Figure from the Doorway of the Town Hall,
Toulon.* 1656. Marble. Life-size. Puget was the only
17th-century French sculptor whose work was almost purely
baroque; all the others were classical to a greater or lesser degree.
This figure shows the characteristics of baroque modelling:
heavy plasticity, deep cutting of the muscles and draperies, and
a complex three-dimensional structure. Although the same
principles had already been seen in Bernini's work, Puget's style
was mainly derived from Pietro da Cortona and his treatment of
emotion recalls Michelangelo. A 'herm' (from the classical god
Hermes) is the technical name for a statue representing the
upper part of the body which ends in a pedestal or plinth out of
which the body appears to spring; such figures were often used
beside doorways to support a balcony, as here.

33. **Simon Vouet.** *The Adoration of the Kings* (detail). *c.*1635–38. Engraving by M. Dorigny, 1640. This is part of a lost ceiling fresco painted by Vouet in the Hôtel Séguier in Paris after his return from Italy in 1627. The decorative style he brought with him was compounded of elements drawn from Roman baroque painting and the art of Veronese. This combination was new to French art and was highly effective for its purpose, although Vouet modified it in keeping with the general trend of the time towards classicism and in deference to classical French taste.

32 (left). **Padre Andrea Pozzo.** *Allegory of the Missionary Work of the Jesuits.* 1691–94. Fresco. Ceiling of the Nave of S. Ignazio, Rome. This late baroque ceiling shows the use of a complex scheme of feigned architecture painted in perspective to create the illusion of recession; the resulting space is much deeper than in the earlier ceiling by Pietro da Cortona (fig. 30). Perspective architecture had been out of fashion in Rome since the beginning of the century, as it had (and has) the disadvantage of only looking right from one spot, immediately below the vanishing point; if the spectator moves away from this spot the illusion 'literally' collapses. But the ceiling was admired on account of the technical skill displayed in it and the method was often copied in Germany and Central Europe, where Pozzo himself settled (in Vienna) in 1702.

packed. Problems of three-dimensional modelling in steep foreshortening have now been mastered with almost as
4   much skill as Rubens could show at this date, although the figures are grouped much less subtly than in the Whitehall Ceiling, which Rubens painted almost at the same time.

Following Cortona, the next stage in Italy was Gaullì's
29 ceiling of the Gesù, in which movement upwards is counterbalanced by the downward movement of the rebel angels tumbling out of the sky. In all these examples it will be noticed how the whole surface is involved. The same is true of easel paintings and altarpieces. The movement of background accessories and the free play of light and shade are integral to the design and contribute to its total effect. One can see this, for example, in Murillo's several variations on
39 the theme of the *Immaculate Conception*, in which the Virgin is borne upwards on billowing clouds and cherubs crowd round her like moths round a lighted candle.

Baroque movement was first introduced into French
33 painting in the late 1620s by Vouet, who had spent the previous dozen years in Italy, just at the time when the new techniques were being developed there. But French taste

was instinctively more classical than elsewhere, and Vouet began to restrain his baroque exuberance soon after settling in France. It was only at the very end of the 17th century, when artists rediscovered Rubens, that the High Baroque in painting was much appreciated in France.

As might be expected, all this had its counterpart in architecture, in the use of curving façades, ingeniously constructed domes, dramatically lit staircases, etc. Baroque buildings of all kinds also tend to be higher in relation to their width than 16th-century buildings, which gives them an effect of upward soaring movement. The great precedent for this in 16th-century architecture was the dome of   4
St Peter's, designed by Michelangelo and built shortly after his death. Its slightly pointed curve gives life and energy to the heavy mass, so that this dome became the almost too perfect model for baroque architects. Movement is also expressed in the handling of architectural detail, as 34,35
in the way in which the vocabulary of classical forms is broken up and different parts of a building are made to flow into one another.

34. **Francesco Borromini.** *Central First Floor Window of the Collegio di Propaganda Fide, Rome.* 1662. Engraving from G. G. de Rossi, *Architettura Civile*, 1692. This engraving dissects a baroque window design. The regular forms and straight-line arrangement of a classical window are 'bent' as far as they will go. The columns stand on a concave curve; the entablature projects forward in the centre in a convex curve, then cuts back to project again at the outer edges at 45 degrees. (The same scheme, writ large, recurs in the façade of S. Carlino, fig. 36.). The triangular pediment is reduced in size and separated from the entablature by a curving form. Such breaches of the classical rules of architecture were strongly condemned by the theorist, Bellori, but it should be noted how calculated they are—they are not merely wayward fantasies.

35. **Mathias Braun.** *Entrance of the Palais Thun-Hohenstein, Prague. c.*1720. The architecture of this doorway shows the distortions of classical patterns and formulae discussed under fig. 34, translated into the bolder, heavier forms of the Central European Baroque. The powerful modelling of the sculptured eagles (by Braun, the leading Bohemian sculptor of the period) may be compared with the figures on the staircase of the Prinz Eugen Stadtpalais in Vienna (fig. 44). The palace was designed by Giovanni Santini.

BAROQUE SPACE

True baroque space, like true baroque movement, was only developed in the 1620s—in architecture, despite some earlier experiments, hardly before the 1630s. Until then painters were preoccupied with normalising space after the arbitrary, irrational distortions characteristic of the later 16th century, and architects with creating practical, hall-like structures in which large numbers of people could assemble. From the point of view of design, buildings were conceived mainly in terms of elevations. Façades, interior walls and arcades in churches were treated in larger, more harmoniously related units and were more richly modelled than before; but structurally their function was to shut off or, in an interior, enclose or sub-divide, space rather than mould, suggest or trap it, as was to be the case after 1630.

The first steps towards a new conception of space in architecture were taken in northern Italy. As early as the mid-16th century, Genoese architects had begun to discover the possibilities of the steeply-rising ground behind the main street on which the most fashionable palaces were built and they constructed staircases of unparalleled in-

genuity at that date, although these were hardly yet Baroque. More significant for the future were the first tentative experiments in the inter-relationship of interior spaces made by the Milanese architect, Ricchino. Ricchino also has the distinction of creating what was probably the earliest façade with a concave curve, that of the Palazzo Elvetico, Milan, which he designed in 1627.

The next stage, which had incalculable consequences for architecture, actually occurred in sculpture. As so often before, Bernini was the originator. His early statues dating from around 1620 are posed in such a manner that they not only produce a powerful psychological and illusionistic effect, but also reach out into space in a visual sense. Air **10** circulates between and around their limbs so that the space, as well as the solid marble, becomes part of the work of art. There is a similar conquest of space and interrelationship of solid and void in the *Baldacchino* in St Peter's. *12*

In pure architecture the projection of forms into space was given both convex and concave forms; convex in the exquisite semi-circular portico attached to the façade of Sta Maria della Pace by Pietro da Cortona (the same *38*

79

36. **Francesco Borromini.** *Façade of S. Carlo alle Quattro Fontane, Rome.* 1665–82. As the date indicates, this façade was only begun long after the rest of the church was finished, but the same architectural principles, expressed by contrasting convex and concave curves, were employed throughout. Note the 45-degree angle of the corner tower, typical of Borromini's contribution to baroque architecture. As the church authorities aptly and approvingly commented: 'Everything is arranged in such a manner that one part supplements the other and the spectator is stimulated to let his eye wander about ceaselessly.'

37. **Francesco Borromini.** *Plan of S. Carlo alle Quattro Fontane ('S. Carlino'), Rome.* 1638–41. From the engraving by J. Sandrart (reversed to show the plan in the same sense as the church). The design of this little church was evolved through a sequence of stages, each based on variations of the possible relationship between circles and two equilateral triangles with a common side. This geometrical approach marked a radical break from the arithmetical approach, based on multiples and sub-divisions of a single unit or module, which was used by Renaissance and classical architects. (Interior illustrated pl. 80).

motive was taken up by Wren and Gibbs in England); concave in the use of façades curving outwards at the ends, especially where the latter are emphasised by towers as in Sta Agnese in Piazza Navona. Towers with open-work **84** upper-storeys of very ingenious design are the most striking examples of the baroque trapping of space. As in Bernini's early sculptures, air and light flow freely around and between the forms—an effect heightened by the contrast of strong light and deep shadow created in Italy by the sunshine. In fact the imaginative use of light and shadow plays as important a part in baroque architecture as it does in baroque painting. Steeples in northern Europe were the equivalent of towers in the south. The steeples in Gibbs's **87** London churches, even more than those of Wren, soar upwards like the painted architecture in illusionistic ceiling decorations, conquering the sky.

A final, untypical, but unsurpassed example, is the relationship of solid and void in Bernini's Colonnade of **4** St Peter's. Although the columns and entablature of this are purely architectural in form, they have a dynamic, sculptural function, powerfully intensified by the sweeping oval curve of the Piazza. Once again space circulates freely between the solid forms; the Colonnade makes a dramatic, definite boundary to the Piazza, yet it is a boundary which is visually and physically penetrated with ease.

Turning to the treatment of interior space, the first important example is Borromini's little church of S. Carlo alle Quattro Fontane (called S. Carlino). The plan of this *36,37* church is basically oval, a form not invented by baroque architects but much used by them and handled with unprecedented inventiveness. Borromini, like Guarini after him, was fascinated by geometry. His method at S. Carlino was to evolve the basic oval of the ground plan by means of intersecting circles produced with a ruler and compasses. He then 'bent' the four quadrants of the oval inwards. The undulating, alternately concave and convex curve which resulted formed the line of the main cornice of the church. At one end of the long axis he placed the entrance, at the other the high altar; side altars filled the concave spaces at each end of the short axis. Above the cornice he used semicircular arches to bring the design back to a simple oval which forms the base of the beautiful coffered dome.

The total effect is thus created by purely architectural means; unlike Bernini, Borromini was only an architect, not a sculptor as well. His architectural forms are of an extreme and daring inventiveness, not only in their interrelationships but also in themselves. They have sharp, cutting edges and, like this architect's plans, are evolved by geometrical means. Nevertheless, the spatial effect of the interior of S. Carlino has analogies with sculpture. The forms, though distinct in themselves, seem to flow into one another and to project and recede in a sculptural way. Visually Borromini's use of geometry is quite unlike that of Renaissance architects. If one stands in a Renaissance building the mathematical basis of the design is so clear

(Continued on page 65)

21. (above). **Peter Paul Rubens.** *Self-portrait with Isabella Brant* (detail). *c.* 1609. Oil on canvas. $70\frac{1}{2} \times 53\frac{1}{2}$ in. (179×136 cm.). Alte Pinakothek, Munich. This portrait of the artist with his first wife was painted about the time of their marriage. With its Early Baroque vivacity, it radiates charm, affection and the success the artist had obtained from his long stay in Italy and with his new position as court painter in his native city, Antwerp.

22. (right). **Gianlorenzo Bernini.** *Portrait-bust of Louis XIV.* 1665. Marble. Life-size. Versailles. This is a portrait of the most powerful personality of the baroque period by that period's greatest sculptor. It was originally intended to stand on a gilt and blue enamel globe, resting on a marble cloth with emblems of victory and virtue in relief. The globe was to be inscribed *Picciola Base*, expressing the idea that the whole world was but a small base for this monarch to stand upon. However, the idealisation of the features in the sculpture itself conveys majesty clearly enough.

23. **Anthony van Dyck.** *Marchesa Balbi.* *c.* 1625. Oil on canvas. 72 × 48 in. (183 × 122 cm.). National Gallery of Art, Washington. (Andrew Mellon Collection).
24. **Diego Velasquez.** *The Infanta Margarita in Blue.* 1659. Oil on canvas. 50 × 42 in. (127 × 107 cm.). Kunsthistorisches Museum, Vienna. These two examples of baroque court portraiture show (left) a lady of the Genoese aris-tocracy and (right) a daughter of Philip IV of Spain. The style of both artists owes something to Titian, but little of his direct influence survives here. Van Dyck is, in fact, more immediately dependent on Rubens, who had visited Genoa twenty years before. However, his primary interest, unlike Rubens's, is in creating an image of elegance, which he does by means of a low viewpoint, making the head and hands small, and choosing sumptuous clothes and a formal yet relaxed pose. Velasquez, more objective, shows a greater awareness of the distinction between the sitter and her role; the Princess stands anxiously in her stiff dress putting on a show of regal authority rather than embodying the reality of it. Velasquez was also a major innovator in the handling of colour and atmosphere, blending the two together in a way not seen again until the time of Impressionism.

25. Frans Hals. *Portrait of a Man.*
c. 1650–52. Oil on canvas. 42 × 33 in.
(106.5 × 84 cm.). Metropolitan Museum
of Art, New York. (Gift of Henry C.
Marquand, 1890).

26. Rembrandt van Rijn. *Lady with a Fan.* 1641. Oil on canvas. 41½ × 33 in. (105 × 84 cm.). Reproduced by Gracious Permission of Her Majesty the Queen. Two examples of Dutch bourgeois portraiture are shown here, both much plainer than the court portraits illustrated on the previous pages and with more emphasis on the faces at the expense of the clothes. Of the two artists, Hals (left) is the more extrovert and robust, treating the sitter in a direct, simple manner, and concentrating his artistic genius in his handling of the brush. With Rembrandt, the brushwork, though finely wrought, is subordinated to the presentation of character, bringing before us a unique physical presence. This presence is insistent and disturbing, yet compellingly 'there'—an effect obtained with the aid of baroque illusionistic devices (the fan and the left thumb overlapping the window-frame), though the portrait is otherwise quite un-baroque.

27. **Rembrandt van Rijn.** *An Old Man Seated.* 1652. Oil on canvas. $43\frac{5}{8} \times 34\frac{5}{8}$ in. (111×88 cm.). Reproduced by courtesy of the Trustees of the National Gallery, London. The introspective possibilities of portraiture are here carried to their ultimate conclusion. The sitter is subsumed in the expression of the artist's idea so that the portrait is essentially a study of contemplation. The open brushwork —softened by glazes at the top left, allowed to remain raw and unfinished at the bottom right—is typical of Rembrandt's later years.

28. **Philippe de Champaigne.** *Omer Talon.* 1649. Oil on canvas. $88\frac{1}{2} \times 63\frac{5}{8}$ in. (224×162 cm.). National Gallery of Art, Washington. (Samuel H. Kress Collection). In contrast to Rembrandt's *An Old Man Seated*, this is a public rather than a private portrait. Although the setting and pose recall the methods of Rubens, the baroque qualities of style are here firmly controlled by the principles of French classicism: clear light, clear colours, sharp outlines and severe restraint in both gesture and expression.

29. (opposite, above). **Peter Lely.** *Two Ladies of the Lake Family. c.* 1660. Oil on canvas. 50 × 70¼ in. (127 × 178 cm.). Reproduced by courtesy of the Trustees of the Tate Gallery, London. An example of baroque portraiture in its English form: the style recalls van Dyck, but the inter-pretation is less refined. The composition has a decorative fullness and breadth characteristic of this phase of the Baroque.

30. (opposite, below). **Nicolas de Largillière.** *Louis XIV and his Heirs. c.* 1710–15. Oil on canvas. 50 × 63 in. (127 × 160 cm.). Reproduced by permission of the Trustees of the Wallace Collection, London. Despite the stiffness of the poses, the scale and the grouping give this work a degree of intimacy previously unknown in court portraiture. The next stage after this was the 'Conversation Piece'.

31. (above). **Francesco Solimena.** *Self-portrait.* Undated. Oil on canvas. 50½ × 44 in. (128 × 112 cm.). Uffizi, Florence. This is an example of Late Baroque portraiture at its most flamboyant. It shows the leader of the Neapolitan School consciously presenting himself as a 'Prince of Painters'. In fact, few portraits of the period make use of so many of the techniques of subject painting: composi-tional inventiveness, bold diagonals and shadow cutting across and into the forms.

32. (opposite). **Godfrey Kneller.** *Matthew Prior.* 1700. Oil on canvas. 54½ × 40⅛ in. (138 × 102 cm.). Reproduced by courtesy of the Trustees of Trinity College, Cambridge. As a work of art this was perhaps an inspired accident, untypical of Kneller and not quite sustained by the technical skill necessary to carry it off (contrast the previous example). Nevertheless, it is one of the most original portraits of its time. Together with certain portrait-busts by Coysevox, it brings into focus the 'new man' of the 18th century: the intellectual—rationalist in spirit, sceptical, urbane and cosmopolitan.

33. (below). **Maurice Quentin de la Tour.** *Self-portrait. c.* 1750–60. Pastel. 25¼ × 21 in. (64 × 53 cm.). Musée de Picardie, Amiens. A more developed example of the same species is shown here; La Tour presents himself as a man of the world rather than as an artist. The pose of the figure is now extremely simple, even casual; vitality is given by the flickering pastel strokes which concentrate attention on the lace at the wrists and throat, on the wig and, above all, on the mocking eyes and mouth.

34. (opposite). **Jean-Marc Nattier.**
Queen Marie Leczinska. Oil on canvas.
53¼ × 38½ in. (135 × 98 cm.). Versailles.
This lively picture is far from being a state
portrait in the conventional Late Baroque
sense. Louis XV's queen is seen in her
town dress and reading the Bible, like any
Scots lady among Allan Ramsay's
clientele. But her wayward glance is less
staid than it would be if she were Scottish
and, discreetly in the background, she is
seen to be sitting on a couch embroidered
with the Royal Lily of France.

35. (above). **Allan Ramsay.** *Mary Atkins
(Mrs Martin).* 1761. Oil on canvas.
50 × 40 in. (127 × 102 cm.). Birmingham
City Art Gallery. Mary Atkins, the sister
of Rear-Admiral Samuel Atkins, married
Admiral William Martin in 1726.
Ramsay's smoothness and daintiness, quite
unlike the robustness of Reynolds or the
bravura of Gainsborough, reflects the
Italian background of his art. But there is
something French in his brushwork and
colour, here reflected in a silvery tonality
that recalls Nattier, as does the rather

nonchalant, almost homely, flavour of the
portrait, which in Britain is hardly
paralleled until Raeburn, another Scot.

36. **Pompeo Batoni.** *John, Marquis of Monthermer. c.* 1760. Oil on canvas. 38¼ × 26¼ in. (97 × 67 cm.). Collection: the Duke of Buccleuch and Queensberry, K. T., Boughton House, Rutland. This portrait, presumably painted in Rome, was earlier attributed to Batoni's chief rival for the English Grand-Touring clientele, Anton Raphael Mengs; but it is suaver and more intimately graceful than the men in Meng's pictures. It combines a sharply characterised, unmistakably English face with a decorative liveliness in the composition only possible in a Continental portrait.

37. **Joshua Reynolds.** *Colonel John Hayes St Leger* (detail). Oil on canvas 93 × 58 in. (236 × 146 cm.). Waddesdon Manor, Bucks., National Trust. Colonel St Leger, who was also painted by Gainsborough, was aide-de-camp to the Prince of Wales and a member of the Hell-fire Club, whose quasi-monastic orgies were held at Medmenham, West Wickham. Reynolds has posed his fiery bachelor coolly and classically enough, but, set as he is upon the edge of an abyss and backed by rolling smoke clouds, he may be seen almost as much in a Satanic light as in the role of a military commander.

that one can measure the distance of each part of the building from the eye. Borromini's approach produces the opposite of spatial clarity. Standing in S. Carlino one is never quite sure how far away its boundaries lie. The undulating curve of the wall and the cornice produces an effect of uncertainty. One is aware of the space as being loosely, elastically contained, not rigidly bounded.

36 The undulating curve of the interior is repeated in the façade designed much later. Here the lower and the upper storeys are contrasted with one another. The lower storey consists of an alternating concave-convex-concave curve, with the entrance door in the convex part. The rhythm of the upper storey runs concave-concave-concave, with a little convex structure projecting from the central concave surface. Further vitality and movement are given by the addition of sculptured figures and an irregular balastrade at the top. The same qualities are to be seen in the façades and interiors of the third great architect of the Roman High Baroque, who was also the painter of the Barberini Ceiling—Pietro da Cortona. Cortona, however, used *38* closely coupled or overlapping pilasters of varying thicknesses to carry the eye smoothly from one part to another, not the irregular forms of Borromini.

The spatial elasticity of the interior of S. Carlino accounts for the parallel with Bernini's architecture, which is otherwise based on more traditional principles. Bernini, *39* too, designed a small oval church, S. Andrea al Quirinale, in which ground plan, cornice and dome follow a continuous, not an undulating curve. But since this curve is itself a non-geometrical form, it creates a similar uncertainty in the definition of the boundary, which is further enhanced by the unusual placing of the chapels either side of the long axis. The same oval curve is used again in the Colonnade of St Peter's. As Bernini described it himself, this Colonnade was designed to symbolise the embracing arms of the Church. Grandeur, power and magnificence are combined with a sensation of almost physical welcome. The effect is the reverse of forbidding—not least because of the warm apricot-coloured stone, slightly rough and pitted in texture, of which the Colonnade, like most Roman buildings, is constructed.

For the next stage in the baroque treatment of space in

38. Thomas Gainsborough. *Mrs R. B. Sheridan. c.* 1783. Oil on canvas. 85 × 58¾ in. (216 × 149 cm.). National Gallery of Art, Washington, (Andrew Mellon Collection). This portrait, which Gainsborough seems to have worked on for over a decade and left not quite finished, may have been exhibited at the Royal Academy in 1783. It shows the wife of the dramatist, Richard Brinsley Sheridan, slightly windswept and *déshabillée* in a landscape, in the painter's most poetic manner. The breeze has brought a becoming blush to Mrs Sheridan's cheeks, harmonising with her dress, while the 'all-over' handwriting of the brushstrokes does not allow the figure to show that detachment from its surroundings which characterises Reynolds's portraits, but produces an atmosphere more truly, if artificially, pastoral. Historically, the portrait belongs at once to the last phase of 18th-century sensibility and the beginnings of Romanticism.

38. Pietro da Cortona. *Façade of Sta Maria della Pace, Rome.* 1656–57. Like Borromini, Cortona used contrasting curves and created deep spaces under arches and porticos to exploit the contrast between cast shadow and the strong Roman sunlight, but he kept the units of his design comparatively distinct and was less capricious in his handling of detail. The result is a more elegant and composed façade than that of S. Carlino. The curved wings do not correspond to the shape of the 15th-century church behind them but conceal an irregular agglomeration of buildings and narrow streets.

39. **Gianlorenzo Bernini.** *Interior of S. Andrea al Quirinale, Rome.* 1658–70. Bernini used a regular oval plan for this church and treated the architectural forms with classical correctness, apart from the cutting back of the pediment over the high altar to take the statue of St Andrew being assumed into heaven. But, although the design has an unsurpassed dignity, comparable to that of the Colonnade of St Peter's (pl. 4) on a much smaller scale, it is as baroque as the interior of S. Carlino (pl. 80), on account of its flexibly treated space. Sculpturally and colouristically it is much richer than S. Carlino.

40. **Guarino Guarini.** *Interior of the Dome of the Chapel of the Holy Shroud, Turin.* 1667–90. The unusual design of the coffering in the pendentives and the geometrical pattern of the ribs in the dome show that Guarini's architecture was derived from Borromini's rather than Bernini's or Pietro da Cortona's. The ribs themselves are transverse, not longitudinal (contrast fig. 39, with longitudinal ribs), and are obtained from the intersection of alternating, diminishing hexagons with the curved surface of the dome. The cross-section of the dome is in the form of a pointed curve, not a hemisphere, producing an almost Gothic effect of upward movement. Small segmental windows are inserted beneath each rib and the whole culminates in a star-shaped 'rose window' at the base of the lantern.

architecture one must return to northern Italy. Here, in Turin, Guarino Guarini, who had visited Rome, created a small group of very important buildings which are even more remarkable than Borromini's in design. Guarini went further than any other Italian architect in abandoning the principles of classicism. In the domes of the Chapel of the Holy Shroud and the Church of S. Lorenzo he created structures of almost Gothic complexity. Instead of the dome being a closed hemispherical surface, as was the case in all previous Italian churches, it becomes in Guarini's hands an open lattice-work structure with ribs crossing the curve like cords, not converging on the centre. The result is the ultimate expression of the baroque conception of space in Italy: here are continuity and elasticity of space, dynamic movement and a glimpse towards the infinite.

Guarini and Borromini were the main sources for Austrian and South German architects. In South Germany the centrally-planned church was less favoured than the more traditional Latin cross plan with a choir, transept and long nave. But ovals were often still the basis of the design, flowing into one another, for example at Vierzehnheiligen by Neumann, in a way that dazzles and astounds. Even more than in Borromini's churches the exact position of the spatial boundary, or rather boundaries, is unclear. Walking about this church, one cannot read its design with the eye. It only appears as a series of fluid interrelated spaces which are barely divided from one another. Curve follows curve both in the walls and in the vaults and these curves are constantly changing as one views them from different angles. Although the details are less fine than in earlier Italian architecture, the spatial effects of South German churches are unparalleled in European art.

Other and in some ways still more dramatic opportunities for the treatment of space in baroque terms were provided by the staircase, since a staircase necessarily involves movement in three dimensions. Admittedly, the

41. **Guarino Guarini.** *Exterior of the same Dome.* The interlocking
stepped effect of the transverse ribs seen from the outside reflects
the complicated structure within. The pagoda-like lantern is
also very unusual. The drum is pierced by six large round-headed
windows. It is typical of Guarini that there is no continuous
cornice separating the drum from the dome but an undulating
line, an idea hinted at, though never precisely carried out,
by Borromini.

42. **Balthasar Neumann.** *Interior of the Church of the Vierzehnheiligen, Northern Bavaria.* 1743–72. The plan of this church is basically a traditional Latin Cross with a nave, transepts and choir, but there is no dome over the crossing and the interior articulation springs from three longitudinal ovals cut by two transverse ovals. The view reproduced here is taken from behind the chancel rail in the choir oval, looking towards the altar of the Fourteen Saints in the central oval, with the south transept, circular in plan, on the left. The inter-penetration of spaces and the manipulation of light which results is the epitome of German baroque architecture. The colouring, some of the detail and the altar tend in style towards the Rococo.

aesthetic limitations as well as possibilities of this feature are determined by its function, for it can never be other than a means of getting from one place to another. But by the late 17th century, the staircase was certainly treated as an end in itself, and in some 18th-century German palaces it is the most prominent feature of the building, with a part of the palace all to itself. (Significantly, there is a word for this in German—*Treppenhaus*, 'stair-house'—which has no exact equivalent in other languages.)

Structurally two basic types of open staircase had already been developed in the 16th century, to begin with in Spain —first, that consisting of four short flights turning through 360 degrees round a square well; second, the type which begins with a single flight up the centre, then divides at the half-landing, returning in two flights to meet at the main landing. The baroque contribution was two-fold: first, to enlarge the ornamental features of the staircase by means of sculptured balustrades and architectural, sculptural or painted decorations on the walls and ceiling; second, to exploit the contrasts in light and shade on the various levels.

The first examples to show the latter are 17th-century Italian staircases (notably the Scala Regia by Bernini in the Vatican), although staircases in Italy were not given particular consideration in the design of the building as a whole and were often fitted into awkward corners. This comparative neglect of the actual siting of the staircase is also often found in French staircases, even at Versailles where the *Escalier des Ambassadeurs* (destroyed in the 18th century) was remarkable chiefly for its grandeur and the magnificence of its decoration, not for its position or the complexity of its design. The full exploitation of the lighting possibilities of the staircase was a development of the late Baroque, for instance the staircase by Fischer von Erlach in *44* the town palace of Prinz Eugen in Vienna. In this example, the visitor comes in through a relatively low, dark vestibule, then ascends the staircase to reach a light open area above. The balustrades are thick and massive with sculptured figures attached to them. In German (as distinct from Austrian) palaces the effect reaches extremes of lightness and spaciousness. The various levels of the stairs themselves are treated more intricately than before with 'quarter landings' on straight flights producing a sort of undulation. Still more attention is given to the space above the stairs. Pommersfelden has galleries in this space; Würzburg has *93* one of the greatest ceiling paintings of the 18th century, Tiepolo's *Olympus with the Four Quarters of the Globe*; Bruchsal, structurally the most ingenious, has an oval plan.

Although other possibilities for the staircase were developed by considering it from the outside of the building, the most characteristic baroque feature of the treat-

ment of outdoor space was the vista. This, too, was a development of the late 17th and early 18th centuries and was associated with the grandiose palace layouts of that period. Earlier one finds more limited town planning schemes, such as the Piazza del Popolo in Rome with its three streets radiating into the city, the centre one of which is framed by two small centrally planned churches by Rainaldi. But really this is not so much a vista as a series of contrasting narrow and open spaces designed to surprise.

Surprise was also the principle of the greatest Roman example in the 17th century, Bernini's approach to St Peter's. This started at the Ponte S. Angelo, with its sculptured figures of angels holding the Instruments of the Passion, and led across the river to the ancient Castel S. Angelo surmounted by an angel with a flaming sword; then the visitor turned left through narrow streets until he reached 4 the Colonnade. Stepping between the columns he was confronted by the immense oval space of the Piazza—a dramatic change from darkness and meanness to magnificence and light. Across the Piazza the space narrowed again for the final stage up the steps to the façade of St Peter's itself. Then once again the visitor had to go through a fairly dark and narrow entrance but, inside, the surprise was repeated again and a new spiritual magnificence was added to the secular splendour left behind. The climax of all this was the 13 *Cathedra Petri* at the west end of the Basilica. Paradoxically,

43. **Francesco de Sanctis.** *The Spanish Steps, Rome.* 1723–25. The contrasting curves of the stairs and balustrades seen in plan produce a sensation of constant variety and surprise (qualities almost of the Rococo) as one walks up. It is as if the architecture had taken charge, controlling one's movements in ways that are hard to resist or understand yet are gentle enough to be delightful rather than disturbing. The church at the top is the SS. Trinità de' Monti, built in the 16th century, with a façade by Maderna.

44. **Johann Bernhard Fischer von Erlach.** *Staircase of the Prinz Eugen Stadtpalais, Vienna.* 1694–8, 1708–11. Next to the stage set, the staircase gave baroque architects more scope for ingenuity than almost anything else. That of the Prinz Eugen Stadtpalais is structurally quite simple, but the transition from the dark ground floor to the well-lit upper floor—a transition dramatised by the thick-set sculpted figures and the massive, very personal architectural detail—is typical of the Central European Baroque.

the greatest vista in baroque Rome is in an interior.

In many ways the outdoor vista, conceived as an avenue with a culminating feature at the end, was more characteristic of northern Europe than the south. Versailles, with its *20* great alleys between the trees and punctuated by lakes and fountains, is the supreme example. It is this which explains one of the otherwise surprising aspects of baroque garden design, namely that it is based on strict symmetry and straight lines, rather than dynamic movement and curves. Vistas are also found in England at this time, for example at Greenwich, where Wren's naval hospital is divided in two, *57* as it were, to reveal a view of the Queen's House across an open space at the far end.

As might be expected, the most striking effects of space in painting occur in ceiling decorations. Perspective architecture was a common method of creating space in these, especially at the beginning and end of the 17th century. (In the middle of the century it went out of fashion.) The first problem for baroque painters was to learn how to treat the ceiling as a single unit of space rather than as a series of separate compartments—a process which took a long time, although one can see the trend towards unification be- *18* ginning with Annibale Carracci's ceiling in the Palazzo *46* Farnese. Guercino's *Aurora* in the Casino Ludovisi marks a further stage in this process, since it has only three compart-

ments, two of which flow into each other by the breaking of the architecture between them.

The management of the whole surface as a single space was first achieved in domes, reviving a method invented in the Renaissance by Correggio. In Pietro da Cortona's ceilings in the Palazzo Pitti, Florence, dating from the 1640s, it was transferred on a small scale to the ceilings of rooms. Finally, from the 1670s onwards it became a commonplace of the treatment of large vaults (the Gesù, S. Ignazio, etc.). *29,32* From this time onwards, too, the space of the ceiling was increasingly merged with the real space of the room. The painted decoration no longer stops at a solid unbroken cornice, but spills over into the real architecture, which itself becomes more and more fanciful. In 18th-century ceil- *84,8* ings the three-dimensional stucco forms, continuing the real architecture, reach up into the ceiling in irregular curving shapes and then recede again like waves on the seashore.

There was a further problem which ceiling painters only solved gradually, namely the extension of the space in depth. The perspective architecture used at the beginning of the 17th century produced this extension automatically, but in the following phase, when painters were employing a looser framework (for example, Guercino in the *Aurora*), the painted space hardly reaches far above the real surface of the ceiling. It was not until the 1660s that the roof was taken off, as it were, and space was extended to infinity. In *32* the late 17th and early 18th centuries effects of unparalleled splendour were achieved by this means. Countless numbers of figures soar upwards, receding as far as the eye can see like stars in the sky.

CONCLUSION

The Baroque is not easy to sum up. It is a centripetal not a centrifugal style, by which one means that all its characteristics are compatible with one another and can be combined in the same work. Such works represent the Baroque in its most typical forms. They are the touchstones, the works by which the Baroque can be recognised and against which others can be tested for the degree to which they approach it. As has been seen, Baroque is very much a style of 'more' or 'less'. A further aspect of this is that not all characteristics need be present at once, yet they all interlock. Illusionism, emotional and psychological appeal, splendour, movement, space, are interrelated qualities; in a sense they are all functions of each other.

This does not mean that the Baroque was not also capable of wide varieties of expression. When examined in detail, Bernini, Borromini, Guercino, Neumann, Rubens, Pietro da Cortona and Tiepolo are seen to be very different from one another; in the Baroque, as in other periods, great masters and major national schools retained their own individuality. But this problem has been deliberately neglected in this chapter, partly because to have treated artists as individual personalities would have made it much more difficult to isolate the characteristics of the Baroque as a style.

Seventeenth-century Classicism

THE CONTRAST BETWEEN CLASSICISM AND THE BAROQUE

Seventeenth-century classicism was opposed to the Baroque in many ways, but the two styles shared their preoccupation with ideal form, their emphasis on eloquence of expression and gesture, and their basic architectural vocabulary. Moreover, classical artists absorbed something of the baroque qualities of luminosity, saturated colour and psychological realism. Sometimes baroque influence went further, as in Sacchi's paintings, Algardi's sculpture and Lebrun's decorative work at Versailles, resulting in the compromise between the two styles known as Baroque-classicism. But over such matters as the appeal to the senses, the fusion of the arts and the treatment of space and movement, there was in principle a marked difference of approach, and this difference became more marked with time.

Almost from the beginning, classicism had the character of a resistance movement against the Baroque. This is not to say that it was simply retrospective; on the contrary, 17th-century classicism was creative, not narrowly imitative or reactionary. Nevertheless, classical artists were more wholehearted in their respect for tradition than their baroque contemporaries. They tried to make their finished works conform to principles derived from ancient and High Renaissance art; they did not merely use prototypes from those sources as raw material to be freely transformed into new designs. Raphael, held in respect by all 17th-century artists (except perhaps Caravaggio), was admired by the classicists barely this side of idolatry. The ideals of the classicists regarding form and composition were clarity, harmony and balance. If they were architects, they believed in preserving the correct use of the orders; if painters, they regarded colour as an adjunct to form, not as equal or superior to it. Beginning with Poussin, if not earlier, they held that art should appeal more to the mind than the senses.

45. **Annibale Carracci.** *The Choice of Hercules.* 1595–97. Oil on canvas. 65¾ × 93¼ in. (167 × 237 cm.). Galleria Nazionale di Capodimonte, Naples. This picture, also known as the *Judgement of Hercules* or *Hercules at the Crossroads*, was originally the centre piece of the frescoed ceiling of the Camerino in the Palazzo Farnese, which Annibale Carracci painted before starting on the ceiling of the Gallery (fig. 18). Its subject is Hercules in the act of deciding between Virtue (left) and Vice, and the picture is thus an allegory of man's power of moral choice. The lucid, harmonious solution to the compositional problem laid the foundations of 17th-century classicism in painting.

46. **Andrea Sacchi.** *Allegory of Divine Wisdom.* 1629–*c.* 1633. Fresco. Palazzo Barberini, Rome. This fresco, painted in a room about half the size of the Gran Salone (fig. 30) and on a much lower ceiling, was a conscious attempt by Sacchi to keep to the classical rules of clarity, simplicity and order in the face of the rising tide of the high Baroque. Cortona's *Triumph of Divine Providence*, painted in the immediately following years, was a high baroque 'answer' to it, and the contrast in methods of approach was debated in the Academy of St Luke.

CLASSICISM IN PAINTING

Few artists other than Poussin quite succeeded in living up to the classical ideals in practice, nor were the differences in attitude between the two sides felt immediately even in theory. The spirit underlying Annibale Carracci's *Choice of Hercules*, painted soon after that artist's move from Bologna to Rome in 1595, was one of eager response to the discovery of the Antique and of Michelangelo's and Raphael's frescoes, but there was no question of its being an anti-baroque manifesto. Even before he left Bologna, Annibale's style had begun to change in a classical direction. When in the following years he painted the Farnese Ceiling, he created a masterpiece that at once revived the traditions of Roman High Renaissance fresco painting and laid the foundations of both the classical and baroque trends in 17th-century art. In this influential work there is a perfect fusion of its various constituent elements and no sense of conflict between them. The sources of the ceiling include motives from Michelangelo, Raphael, Titian and the Antique; its style combines something of the clarity and firmness of outline characteristic of classicism with a decorative richness and exuberance suggestive of the Baroque. Maderna's façade of Sta Susanna combines classical and baroque qualities in architecture in the same way.

However, among the next generation of painters working in Rome, especially the Carracci pupils, there was a perceptible classical reaction. The classical implications of the Farnese Ceiling were followed up and its baroque ones ignored. By the early 1620s a contrast between the two styles was becoming apparent. It is possible that Guercino, who had come to Rome in 1621 with one of the most advanced baroque styles in Italy, was advised by his patron's learned secretary, Monsignor Agucchi, to change to a more classical manner, and acted on the advice, although he never became a fully classical artist. In the same years Guido Reni was cultivating another quality on which classical artists and theorists laid stress—ideal beauty. Although Reni's paintings are baroque in sentiment and handling, they exhibit a classical emphasis on the perfect human body which with Reni was almost an obsession.

In 1629 Andrea Sacchi began a ceiling in the Barberini

46 Palace, the *Allegory of Divine Wisdom*, that likewise had certain baroque features but was composed, unillusionistically and with relatively few figures, on classical lines. Al-
30 most immediately it was 'answered' by another, larger ceiling in the same palace, the *Triumph of Divine Providence* by Pietro da Cortona—the first great masterpiece of Roman high baroque painting—which has already been discussed.

Despite Cortona's resounding triumph for the Baroque in the 1630s it began to seem in the next decade as if classicism was gaining ground. Compositions became clearer, figures were more spaced out and were reduced in number and foreshortenings were tamed. Even the most baroque artists, including Bernini, succumbed to these tendencies to some extent. Nor was the change of mood confined only to Italian artists or to artists who belonged in the broadest sense to the classical tradition; it was felt throughout all European art. In Rome a small group of painters, of whom the best known was Sassoferrato, went back to the 15th century for inspiration, producing a sort of Pre-Raphaelitism looking like that of the German Nazarenes of the early 19th century. Elsewhere in Italy the Baroque was less drastically modified. Outside Italy a new calm, simplicity and order permeated the works of such otherwise unclassical painters as Velasquez and Rembrandt at this time. Dutch landscape, *genre* and still-life painting were similarly affected although they were quite unconnected with classicism in the formal sense. A spirit of *recueillement* set in; it was as if painters all over Europe had paused for thought.

The greatest representative of this phase was Nicolas Poussin. Poussin made classicism into a serious, expressive and deeply moving pictorial language as perhaps never before or since; to find anything comparable to his art one has to go back to the Parthenon Frieze, which he never knew. By temperament he was an admirer of the Venetians
43 and his early works glow with colour and a poignant sensual charm, but he gradually purified his style of these attractions in the interests of clarity and intellectual precision. By 1642 he could write: 'My nature leads me to seek out and cherish things that are well ordered, shunning confusion which is as contrary and menacing to me as dark shadows are to the light of day.' At the same time he was
47 making careful studies after ancient statues and bas-reliefs, and, when he came to compose his paintings, would sometimes approximate the final version of a form more closely to the classical original than it had been in the first sketch.

Poussin's saturation in the moral attitudes and visible remains of the ancient world gave a mood of austere gravity to his mature and late works. He took great care to arrange his figures in such a way that every gesture and expression would tell. Nothing was allowed to be either superfluous or
10 unclear. Standing in front of such a work as the *Sacrament of Confirmation*, we are the witnesses of some high, solemn ceremony (our role is very definitely that of observers rather than participants). All movement in the picture is restrained, gestures are restricted to lines parallel to the surface and space is plotted with mathematical exactness.

47. **Nicolas Poussin.** *Studies after the Antique. c.* 1645–50. Pen and brown wash. $12\frac{1}{2} \times 8\frac{1}{4}$ in. (32×21 cm.). Musée de Valenciennes. Ancient Greek and Roman sculpture was naturally of great interest to Poussin (though perhaps no more so than to Rubens) and he made many drawings after it, particularly after sculptured reliefs. He used the knowledge so gained not only for details and accessories but also—and here he differed from Rubens and other baroque artists—as a guide to the finished appearance of his paintings as a whole.

There is no blending of forms and no extravagance of emotion, as there would be in baroque art. Everything is governed by the intelligence and appeals to the intelligence for its appreciation. Yet for all the deliberation of Poussin's art, for all his reputedly conscious attempt to suppress feeling, his paintings are hardly ever cold or lifeless. Classicism in the hands of a great artist does not mean emotional sterility but an emotional tension—a tension developed between the imaginative force of the informing idea and the strict discipline of the means used to control and express it.

IDEAL LANDSCAPE PAINTING

Another type of classical painting which flourished in Rome at this time was ideal landscape, of which Poussin 62 himself was one of the greatest masters. The sources of ideal landscape can be traced back to the Renaissance but the founder of the tradition in its 17th-century form was Annibale Carracci. Many of its practitioners were foreign- 58

48. **Gaspard Dughet,** also called **Gaspard Poussin.** *Classical Landscape. c.* 1650–70. Engraving by Châtelain, 1741. Gaspard Dughet was the third great classical landscape painter working in Rome in the 17th century, ranking next after Nicolas Poussin (who married his sister) and Claude Lorrain. His work chiefly shows the influence of his brother-in-law but his treatment of nature is less severely mathematical and he occasionally introduces a pastoral note or lighting effect reminiscent of Claude. All the conventions of classical landscape are set out in in this engraving in almost 'text-book' fashion.

49. **Claude Lorrain.** *Study of Trees. c.* 1640. Brown wash and black chalk on pale ochre-tinted paper. $9\frac{7}{8} \times 7\frac{1}{8}$ in. (25 × 18.2 cm.). British Museum, London. It was by making studies from nature such as this that Claude acquired the knowledge of tree forms, atmosphere and light that he used in his paintings. The early *Landscape with a Goatherd* (pl. 57) is only slightly more idealised than this drawing and is similar to it in feeling. But the drawing itself is already simplified and idealised to some extent and is a form of visual poetry, not a literal record, and is a work of art in its own right.

ers, of whom Poussin and Claude Lorrain were French, Poussin's brother-in-law, Gaspard Dughet, was half-French and Adam Elsheimer was German. (Although the latter was not strictly an ideal landscape painter in the classical sense he belonged to the tradition by association.)

48
59

The essence of ideal landscape painting lay in the conjunction of two things: drawing from nature in the countryside round Rome (the Campagna), and the use of a certain set of pictorial conventions. These conventions might be applied in a variety of ways and with greater or lesser degrees of elaboration but, reduced to their simplest, they would produce a landscape composed as follows. The foreground would consist of a more or less flat plain, often with a stream running through it. To one side and generally rising the whole height of the picture there would be a tree or group of trees with spreading foliage, seen partly in shadow and silhouetted against the sky. On or towards the other side and set further back in space there would be a second group of smaller trees, balancing the first, often near some rising ground surmounted by a classical building. The distance would consist of further flat ground, possibly with a winding river, and, on the horizon, a faint line of hills or a view of the sea. Two other essential ingredients were figures, often drawn from some story in the Bible or classical mythology, and an all-pervading light. This light was in many ways the key feature of the system. Elsheimer

was the first to discover the poetic possibilities of light enveloping the whole of a landscape composition, but it was Claude who treated it with a subtlety unparalleled in his time and not surpassed in representational accuracy until the days of Impressionism.

All the natural features of ideal landscape, including the light, were inspired by the Roman countryside, and many of the drawings made by artists from nature—especially the drawings of Claude—are of great beauty in their own right. Nevertheless, the end which the artists had in mind in their paintings was not to represent nature as they saw it, but to treat it in an idealised way. They transformed the Campagna into a region of enchantment and delight, conceiving it in imagination as it might have been in a remote Golden Age. As in Poussin's figure paintings, the elements of the landscape were carefully selected and related to each other according to principles of clarity and order.

49

Despite the consistency of the conventions employed, ideal landscape was neither a monotonous nor stereotyped form of art. It was capable of almost infinitely varied nuances of interpretation and several major differences of emphasis. Poussin, as might be expected, stressed logic and clarity, preferring strong sunlight and sharp outlines and applying to the confusion of external nature the same principles of mathematical order which he introduced into his figure compositions. Claude, on the other hand, chose a

62

50,57

50. **Claude Lorrain.** *Study for ' Landscape with the Nymph Egeria'.* 1663. Pen, shades of brown and grey-brown wash and white body-colour. 7 × 9½ in. (17.8 × 24 cm.). English Private Collection. This drawing is a study for a painting, the central part of which is reproduced on pl. 61. Such highly finished drawings were as typical of Claude's later years as the nature studies were of his earlier, although he produced examples of both types of drawing throughout his life. This drawing is like a painting in all but medium.

softer light, often the light of morning or evening rather than midday, and created a more idyllic mood. Sometimes he actually represented the sun in his paintings using it as the source of light for the first time in the history of art. If the sun is not visible, the light still emanates from an area of the sky just above the horizon, filling the whole composition with its radiance and linking foreground and background in a continuous spatial unity.

Whether this discovery of the poetry of light is a specifically classical characteristic is an open question; in some of Claude's early harbour scenes the light has a splendour and theatrical intensity that are more typical of the Baroque. But pervasive, poetic light is generally associated with moods of serenity and calm. In Italy such moods always occur in landscapes that are also classical in form; in Holland in the 1640s and 1650s they occur in landscapes that are classical only in spirit but, in keeping with the climate, a subdued, misty light plays almost as central a part in the landscapes of van Goyen or Cuyp as it does in those of Claude.

Although Poussin and Claude spent their working lives in Rome, they were typically French artists, in that it was in France rather than Italy or any other country that classicism found its most constant expression. After Poussin's death in 1665, the Baroque was once more in the ascendant in Italy and such artists as attempted to resist it in the name

of classicism were pulled further than before in a baroque direction. It was as if the whole conflict had shifted its ground and was taking place more on baroque terms. It was not until the mid-18th century that classical ideals were revived in Italian painting and even then the revival was largely confined to Rome. In the liveliest centre of Italian painting at that time, Venice, classicism was hardly felt at all except perhaps as a quietening of style in the last works of Tiepolo. Nor did classicism find much echo in any of the arts in Spain, Germany and Central Europe until the advent of Neo-Classicism in the last quarter of the 18th century.

The classical bias of French painting was naturally due in part to Poussin's influence, but it can also be found before that influence became effective (in the early 1640s), at least as an attitude of mind if not yet as a desire to imitate classical antiquity. The French Caravaggesque painter Georges de la Tour, who worked in the small town of Lunéville in Lorraine, differed from Caravaggio himself in precisely this way. He used all the Caravaggesque devices of close-up, surface realism and strong tone contrasts but assembled his compositions with a new economy and restraint. Even in harrowing scenes such as the *St Sebastian tended by St Irene* there is no movement, no violence and very little blood; there is only the motionless body of the saint on the ground, a single arrow and the silent figures of St Irene and her serving women coming to tend him. Another instance of the way the inherently classical tendency of French taste could influence an unclassical type of art can be seen in the work of the *genre* painter, Louis le Nain. The baroque decorative artist, Simon Vouet, also began to modify his style in a classical direction after his arrival in Paris from Rome in 1627, although he might have done this in any case in conformity with the general trend of European art.

From 1640–42 Poussin himself stayed in Paris, having been summoned there by the King to decorate the long gallery of the Louvre. The visit itself was a failure and Poussin returned to Rome in disgust but his influence was implanted in Paris to become a permanent feature of French art. Under that influence, which was reinforced by that of Lebrun who had been his pupil in Rome, the French Academy evolved the theoretical doctrines which fixed the rules of classicism for the next hundred years.

CLASSICISM IN SCULPTURE

In view of the eagerness with which ancient sculpture had been studied ever since the Renaissance, one might expect there to have been as strong a classical current in 17th-century sculpture as there was in painting. In fact this was not so and it is difficult to find a reason for it other than the accident that no classical sculptor was born who was anywhere near talented enough to challenge the supremacy of Bernini. Until the middle of the century there were few good sculptors in France and, by European standards, none at all in England or Holland. In Flanders, apart from

51. **Georges de la Tour.** *St Sebastian tended by St Irene. c.* 1650.
Oil on canvas. 63 × 50¾ in. (160 × 129 cm.). National Museum,
Berlin-Dahlem. This classical interpretation of Caravaggio
should be compared with Ribera's baroque interpretation shown
in fig. 24. The caption to that work also contains a note on the
subject-matter. The characteristic of Georges de la Tour's
last period, to which this painting belongs, is an extreme
simplification of form, to the point where the figures seem
almost as if made of carved wood.

Duquesnoy, who spent most of his career in Rome, the most important sculptor was Rubens, although his designs for work in this medium were mostly carried out by assistants. In Spain there were two main types of sculpture, neither of which was classical: first, decorative sculpture used in conjunction with architecture and of no great interest in itself; second, carved and coloured wooden sculpture, which will be discussed under the heading of realism in the next chapter.

The two important sculptors working in Rome who achieved real individuality and showed this in at least some classical ways were Duquesnoy, who was Flemish by birth, and Algardi. Duquesnoy was a friend of Poussin and his statue of Sta Susanna in Sta Maria di Loreto was conceived *52* on consciously classical lines. The figure stands clear in a niche with one leg taking most of the weight and the other leg trailing in the manner of the Antique. The folds of the drapery fall naturally according to the laws of gravity and are not agitated by the emotion of the figure as they would be in a statue by Bernini. One hand points across the body to the altar; the head turned in the other direction looks towards the congregation (or rather, this was the original arrangement—the figure now stands on the wrong side of

52. **François Duquesnoy.** *Sta Susanna. c.* 1626–33. Marble. Life-size. Sta Maria di Loreto, Rome. This is to Bernini's sculpture as Sacchi's paintings are to Pietro da Cortona's (figs. 46, 30). The figure is at present placed on the wrong side of the church; it originally stood so that the saint pointed in the direction of the altar and looked towards the congregation, thus creating a baroque relationship between the real world, the world of art and the world of spirit. Otherwise the statue shows all the marks of a consciously considered classicism.

53. **Alessandro Algardi.** *Portrait Bust of Cardinal Domenico Ginnasi. c.* 1630–40. Marble. Life-size. Galleria Borghese, Rome. The restrained treatment of the drapery and clean silhouette of this bust invite comparison and contrast with the dynamic, aggressively baroque bust of Louis XIV by Bernini (pl. 22). At the same time, the face is more expressive and realistic than that of Duquesnoy's Sta Susanna, giving Algardi a place mid-way between his two great contemporaries in the baroque-classical spectrum.

54. **François Girardon.** *The Grotto of Thetis.* 1666–73. Marble. Life-size figures. Engraving by J. Edelinck, 1678. This rather disagreeable engraving shows the original arrangement and setting of the group which now stands, partly altered, in a naturalistic grotto made in the 18th century in the gardens at Versailles. Girardon clearly conceived the work as a three-dimensional picture, or as a sculpture in very high relief, giving it a style and composition that proclaimed its debt to Poussin. The subject is Apollo tended by Nymphs in the Grotto of Thetis after his work (of driving the chariot of the sun across the sky) was completed at the end of the day.

the church). The features and hair are modelled according to classical conventions, the eyeballs are unmarked and the expression is one of thoughtful tenderness. The only baroque characteristic is the pronounced curve of the figure outwards into the space of the church—a characteristic only visible from the side.

In his subsequent work Duquesnoy was more strongly affected by baroque influences. Algardi could not avoid these either and he had in addition a baroque feeling for the plasticity of form and surface. His carving of a head is in some ways even more subtle than Bernini's, though it is more restrained. From a naturalistic point of view he could create the illusion of hair, eyes, ears and lined and wrinkled skin, all in white marble, with incredible skill. Like Duquesnoy he held feeling and movement in check in most of his works and kept the outlines of his figures clear.

Sculpture was less important in France than either painting or architecture and, with one or two striking exceptions, was chiefly used in a decorative context. In contrast to Italy, classicism is the rule, baroque the exception —an exception which, however, tended gradually to modify the rule towards the end of the period. Most French sculptors spent a period of training in Italy, where they studied the Antique as well as contemporary Italian art. What they carefully avoided was the influence of Bernini, who was less important for French sculpture than was Poussin, even though the latter was a painter. Moreover there was often an association between French sculptors and contemporary architects and painters: Sarrazin worked with Vouet and Mansart, and Girardon and Coysevox were part of the team that created Versailles.

Sculptors at Versailles were employed both inside the Château and for making decorative statuary and fountains for the gardens. The general atmosphere of Versailles and the context in which all this sculpture was seen—not framed in isolation, but combined with other arts and with nature —inevitably gave it a certain baroque flavour. Indeed the sculpture is among the chief sources of the decorative splendour of Versailles. Its function is to attract, impress and charm the eye. Yet in detail the work of Girardon at any rate—Coysevox's rather less so—was classical in style. The former's masterpiece, the *Grotto of Thetis*, is directly reminiscent of Poussin in composition and seems to show the artist almost deliberately resisting the baroque possibili-

55. **Antoine Coysevox.** *Tomb of Cardinal Mazarin.* 1689–93. Marble and bronze. Life-size figures. Louvre, Paris. This tomb may be compared with Bernini's of Pope Alexander VII (pl. 85). But whereas the Italian master used all the available baroque devices of illusionism, rhetoric and a rich and varied colour scheme, Coysevox built up his composition into an almost symmetrical pyramid, keeping to the most economical means of expression and confining his materials to white marble and dark bronze.

ties of the commission he was given. An engraving made before the setting was altered shows that Girardon isolated the figure group as far as possible from the niche containing it, whereas a baroque sculptor would have interrelated the two. The figures themselves are treated as separate units with clear spaces between them; their gestures and expressions are restrained and their movements kept parallel to the surface. The same restraint, parallelism and detachment can be seen in Coysevox's tomb of Cardinal Mazarin (1689–93). The sculptural and architectural components of the monument are kept sharply distinct from one another with the three separate allegorical figures on one level and the effigy of the Cardinal, accompanied by only a single *putto* with a *fasces*, on the other.

Coysevox was instinctively a more baroque sculptor than Girardon but he too moderated the natural exuberance of his style according to the classical dictates of economy and restraint. Even though his equestrian statue of *Fame*, made originally for the gardens of Louis XIV's Château at Marly, has a certain baroque panache and is superbly decorative, it has none of the rippling three-dimensional movement of a horse and rider by Bernini. The same quali-

ties can be seen in Coysevox's portrait busts, a field in which he was one of the finest masters of the 17th century. In place of Bernini's dynamic twist of the head in the *Louis XIV*, which makes it seem as if the figure had just turned imperiously to speak or look at something, Coysevox uses only a slight contrast between the line of the body and that of the head. The expression is pensive, the drapery merely a formal suggestion of a classical robe and the silhouette of the bust simple and regular.

CLASSICISM IN ARCHITECTURE

There seems to have been little interest in classical architectural theory in Rome before the last quarter of the 17th century. Perhaps partly for this reason, and partly once more because of the accident of birth, the classical current in the Italian architecture of the period was even weaker than that in sculpture. The high baroque architects, Bernini, Borromini, Pietro da Cortona and Guarini, had no classical contemporaries who would deserve mention in so short a book as this. At the beginning of the century, Maderna's work showed an unselfconscious blend of classical and early baroque elements that corresponded

56. Carlo Fontana. *Façade of S. Marcello al Corso, Rome.*
1682–83. Fontana was the leading architect in Rome in the last
twenty years of the 17th century, after the deaths of Bernini,
Borromini and Pietro da Cortona. Though partly echoing
Cortona's façade of Sta Maria della Pace (fig. 38), S. Marcello
marks a return to the dignified restraint of Maderna's
Sta Susanna of 1597–1603 (pl. 79). Its architect was skilful,
suave and industrious rather than great; among his north
European admirers and pupils were Hildebrandt, Pöppelmann
and Gibbs.

broadly to Annibale Carracci's style in painting. It was
only in the 1680s that the classical current became strong
enough to create an anti-baroque reaction, and even that
very partial. Perhaps Bellori's attacks on Borromini in the
Idea (1672) encouraged the new trend, by recalling Italian
architects to their classical and Renaissance heritage. On
the other hand, the Roman atmosphere of the time was so
saturated with baroque feeling that this could not be
escaped either. The chief exponent of the resulting 'baroque-classical' compromise was Carlo Fontana, whose
façade of S. Marcello combines a more or less classically   *56*
correct use of the orders with a baroque context. The
concave curve of the façade and the continuous upward
movement joining the lower and upper storeys are baroque;
yet the façade has no contrasting convex curve and the
various formal units—columns, pilasters, pediments and
so on—are handled in such a way as to be easily read.

The fullest and most perfect realisation of classicism in
architecture, even more than in painting and sculpture,
occurred in France. The classical movement there began as
early as the second decade of the century with the work of
the first French architect of the period, Salomon de Brosse.
De Brosse's chief innovations were, first, as strict as possible
a regard for symmetry, so that the sides as well as the back
and front of a free-standing building could be read as a
balanced unit and, second, conception of the whole
building in terms of mass rather than surface ornament.

The second of these innovations was to be particularly
important to the leading French architect of the next
generation, François Mansart. All Mansart's buildings are
conceived three-dimensionally and depend for their effect
on the exact relationship of the various parts. This harmony of parts is the essence of classical architecture. In the
Orléans Wing at Blois, the regularly repeated windows   *58*
and restrained pilaster order on all three storeys articulate
an exterior divided into four main parts, i.e. a *corps-de-
logis* (body of the building), a central frontispiece slightly
raised on the upper storey, and two short projecting wings.
Even the traditional French pitched roof, modified by
Mansart into the hipped form which bears his name, is
made part of the scheme by the adjustment of its proportions to the strictly classical frontage below. Sharp edges
and angles impart precision to the design. There is no

(Continued on page 97)

39. (opposite). **Bartolomé Murillo.** *The Immaculate
Conception. c.* 1660. Oil on canvas. 81 × 56¾ in. (206 × 144 cm.).
Prado, Madrid. Although this work, at one time in the
Escorial, does not sum up all possible kinds of 17th-century
religious feeling, let alone of baroque pictorial style, its theme
and manner of presentation can both be considered
representative. First, it exemplifies the cult of the Blessed Virgin
(specifically, her innocence of Original Sin); second, it is
emotional in treatment; third, its composition consists of a
shifting, irregular pattern of light, shade and colour to which all
elements of the picture contribute and which merge into one
another.

40. (above). **Michelangelo da Caravaggio.** *David with the Head of Goliath.*
c. 1605. Oil on canvas. 49¼ × 39¾ in.
(125 × 101 cm.). Borghese Gallery, Rome.
41. (opposite). **Guido Reni.** *St John the
Baptist. c.* 1640. Oil on canvas.
88½ × 63¾ in. (225 × 162 cm.).
Dulwich College Picture Gallery, London,

reproduced by permission of the
Governors of Dulwich College. These two
interpretations of the young male nude
from the Early Baroque period are both
simplified in form and highly evocative in
feeling. Caravaggio's *David* is harsh,
tragic, and ambiguous in its implications.

The realistically painted surfaces, the
darkness, the sword and the violence are
the hallmarks of his art, and the head of
Goliath is modelled on his own features.
Reni's *St John*, by contrast, shows an
idealised, unequivocally beautiful youth,
sensuous without being sensual.

42. (left). **Pietro da Cortona.** *The Rape of the Sabine Women* (detail). *c.* 1629. Oil on canvas. 108 × 166½ in. (275 × 423 cm.). Capitoline Gallery, Rome. Cortona's *Rape of the Sabine Women* is one of the first secular examples of Roman High Baroque painting. Note the greater complexity of the forms compared with Caravaggio's or Guido Reni's and the introduction of a new quality of dynamic physical energy— appropriate to the subject, no doubt, but typical of Cortona's work. His figure style has a plastic richness comparable with Bernini's in sculpture.

43. (opposite). **Nicolas Poussin.** *Diana and Endymion* (detail). *c.* 1631. Oil on canvas. 47½ × 66 in. (121 × 168 cm.). Detroit Institute of Arts. This is an early, comparatively baroque work by the greatest master of 17th-century classicism (see plate 10). The fluid brushwork, the sense of movement and the glowing colours are qualities comparable to those in the picture on the left, painted in the same city, Rome, at almost exactly the same moment. Poussin's picture is, however, more lyrical and refined both in feeling and in style (it is one of the most poetic subject pictures in 17th-century art) and already shows something in the poses and facial expressions of the pure classi- cism he was to develop a few years later.

44. (left). **Mattia Preti.** *Modello for the Plague Fresco.* 1656. Oil on canvas. 50¾ × 30¼ in. (129 × 77 cm.). Galleria Nazionale di Capodimonte, Naples. This is a finished oil study for one of a series of frescoes (now destroyed) painted by Preti to be displayed over the gates of Naples in commemoration of the plague which struck the city in 1656. Saints intercede with the Madonna on behalf of the victims seen below, while an angel with a flaming sword and a scourge is sent to drive out the infection. Stylistically, the painting exemplifies the Neapolitan High Baroque: daring, broadly treated, heavily shadowed and with a streak of the morbid in its emotional content.

45. (opposite). **Giambattista Tiepolo.** *The Trinity appearing to St Clement* (?). *c.* 1730–35. Oil on canvas. 27¼ × 21¾ in. (69 × 55 cm.). Reproduced by courtesy of the Trustees of the National Gallery, London. The greater airiness, brilliance of colour and virtuosity of handling displayed in this painting contrast with the relatively crude brushwork and raw emotions shown in the work opposite. Tiepolo's picture was painted as a preparatory study for an altarpiece in the Chapel of the Nuns of Nôtre Dame at Nymphenburg, now in the Munich Gallery, but it is so highly finished that it might almost be a smaller, alternative version of the final work.

46. (following pages). **Guercino.**
Aurora (detail). 1621–23. Fresco. Casino
Ludovisi, Rome. This fresco, painted for
the nephew of Pope Gregory XV on a
coved ceiling in the principal room of the
Casino Ludovisi, represents the transition
in style from the Early to the High Baroque
in Rome. The relatively clear outlines and
simplified forms are typical of the former
phase, while the dynamic movement,
dark colours and strong tone contrasts
(never before seen in a fresco) belong to
the latter. Aurora (the Dawn) in her
chariot charges across the sky scattering
flowers. Outside the detail shown here are
to be seen her lover, Tithonus, whom she
has just left, and the fleeing figure of Night,
whom she chases away.

47. (opposite). **Peter Paul Rubens.** *The Garden of Love* (detail). *c.* 1630. Oil on canvas. 50 × 68 in. (127 × 173 cm.). Waddesdon Manor, Bucks., National Trust. These are three figures from the right side of the first version of Ruben's *Garden of Love*; a later, more famous version is in the Prado. The subject is a *fête galante*, in which a number of elegantly dressed couples, accompanied by Cupids and watched over by a statue of Venus, are seen talking to each other and embracing on the terrace of a house. Such paintings by Rubens strongly influenced Watteau and were used, transformed from a baroque into a rococo idiom, as the basis of much French 18th-century painting (see plates 19 and 67).

48. (below). **Rembrandt van Rijn.** *Bathsheba.* 1654. Oil on canvas. 56 × 56 in. (142 × 142 cm.). Louvre, Paris. The courtly idyll of profane love shown opposite may be contrasted with this painting by Rembrandt—one of the noblest, most poignant and psychologically most realistic treatments of sex in the history of art. The whole composition, not just the figure of Bathsheba herself, seems to communicate her mixed feelings of desire and foreboding, as she holds, but no longer reads, King David's letter. The nude is one of the few examples in Rembrandt's art in which he does not denigrate the female body, hence it is perhaps not fortuitous that he modelled the pose on an engraving after a classical relief.

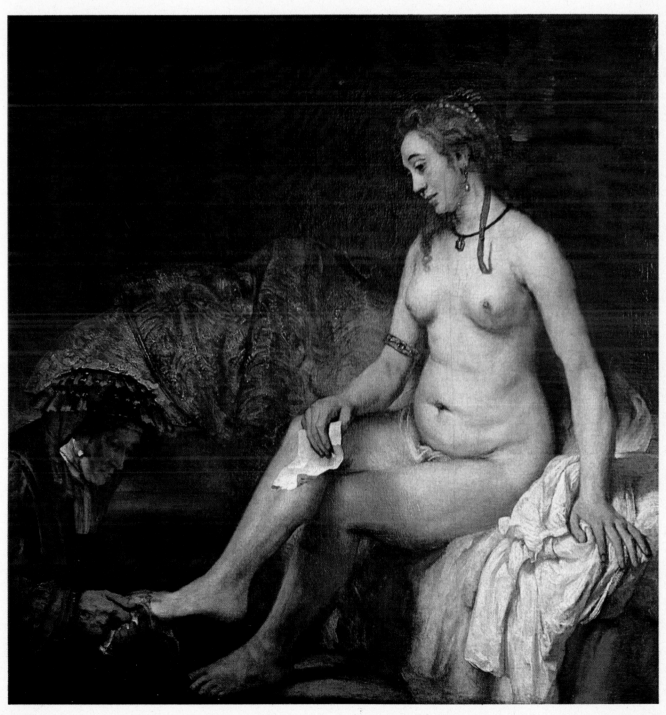

49. (above). **Jan Steen.** *The World Upside-Down. c.* 1663. Oil on canvas. 41¼ × 57 in. (105 × 145 cm.). Kunsthistorisches Museum, Vienna. The canvas is inscribed *In Weelde Siet Toe* (When you lead the sweet life, be prudent). This use of a tag or proverb is typical of Steen, who was Holland's greatest painter of popular life. But while he intended a warning, there is no doubt that he enjoyed depicting scenes of which he ostensibly disapproved.

50. (right). **Louis Le Nain.** *The Farm Waggon.* 1641. Oil on canvas. 22 × 28½ in. (56 × 72 cm.). Louvre, Paris. Scenes of low life are rarer in French 17th-century painting than in Dutch, though they occur quite frequently in French graphic art. Louis Le Nain was the most sensitive painter in France of this type of theme. He treats the peasants who were his subjects with great sympathy, even dignity, but is completely reserved, detached and objective about them. They seem to stare back at the spectator with an air of Stoic contemplation. There is a certain parallel with Poussin's attitude, though style and subject-matter are quite different.

51. **Jan Vermeer.** *A Lady and Gentleman at the Virginals. c.* 1660. Oil on canvas. 28½ × 24½ in. (72 × 62 cm.). Reproduced by Gracious Permission of Her Majesty the Queen, Buckingham Palace. Vermeer's still, silent, contemplative art represents a sort of classicism without the influence of classical antiquity. He was the great master of Dutch bourgeois genre painting. Ladies and gentlemen, separately or together, occupy his carefully furnished interiors, talking, reading or writing letters, drinking, pouring milk or, as here, playing music. The instrument is a virginals, an earlier version of the harpsichord, its lid inscribed *Musica Letitiae Comes Medicina Dolor[um]* (Music is the companion of joy, the balm of sorrows). The painting also exemplifies Vermeer's controlled rendering of daylight and, on the back wall, his fondness for a geometrical pattern of rectangles.

52. **François Boucher.** *Cupid a Captive.*
1754. Oil on canvas. 64½ × 32⅝ in.
(164 × 83 cm.). Reproduced by permission of the Trustees of the Wallace
Collection, London. This panel, which
shows Cupid held captive by the Three
Graces, was painted, with three others, as
a decoration for the boudoir of Boucher's
friend and protectress, Madame de
Pompadour. It was through the influence
of Louis XV's most celebrated mistress
that Boucher became First Painter to the
King in 1765, and his weightless and
pretty treatment of traditional classical
subjects is one aspect of the *style Pompadour*.
The fleshier aspects of Rubens's vigour
have become the rhetoric of titillation, and
only in the immense *Rising* and *Setting
Suns*, also in the Wallace Collection, does
Boucher come close to Rubens's grandeur.
Here he creates a decorative, asymetrical
soufflé of forms very similar in feeling to a
figure group in porcelain of the period,
or to a silver table ornament by Meisso-
nier.

53. (right). **Jean-Baptiste Chardin.** *The Morning Toilet. c.* 1740. Oil on canvas. 19¼ × 15¼ in. (49 × 39 cm.). National-museum, Stockholm. From the ingredients of childhood aping adulthood, and adult-hood wishing to mirror itself in childhood, Chardin has managed to produce a remarkably unsentimental picture, replying on a simple, pyramidal composi-tion, symmetrical and balanced in spite of the casual lopping of the furniture at the edges, and allowing himself only the coffee-pot in the foreground to demon-strate his astounding facility with still life. Chardin undoubtedly knew the work of some of the Dutch 17th-century genre painters (though not Vermeer), but his art has a purely pictorial and aesthetic beauty, independent not so much of subject-matter as of illusionism, to which theirs still clings.

54. (below). **J.-C. Duplessis.** *Inkstand. c.* 1770. Apple-green and white Sèvres porcelain, with gilding. 7 in. long. (17.8 cm.). Reproduced by permission of the Trustees of the Wallace Collection, London. The style of this piece seems to be a little old-fashioned for 1770, since the more flamboyant Louis XVI style had already superseded the rich but simpler fashion associated with the reign of Louis XV, whose portrait is on the cameo and who presented the inkstand to his daughter-in-law, Marie Antoinette. The painted Cupid decorations by Falot indicate that this is a marriage gift, and the terrestrial and celestial globes flanking the crown of France may be intended as a rather fanciful political emblem.

55. (right). **William Hogarth.** *The Countess's Dressing-Room. c.* 1743. Oil on canvas. 27¾ × 35¾ in. (70.5 × 91 cm.). Reproduced by courtesy of the Trustees of the National Gallery, London. This is the fourth scene from Hogarth's most famous series of 'Modern Moral Subjects', the *Marriage à la Mode*. The style is realistic, though the stilted poses and artificial movements suggest the contrived realism of the theatre; but Hogarth's intention is didactic and satirical, not documentary, and the whole series is an attack on immorality in high life and on the follies of marrying for money.

sensation of movement as there would be in a baroque building; instead, there are one or two graceful added touches—the ornamental motives in the two upper storeys of the frontispiece and the curved colonnades in the angles on the ground floor—which enliven what might otherwise be a forbiddingly severe front. At no point do these touches obscure the structure of the building, which is lucid, monumental and scrupulously finished in detail.

Mansart's classical style has rightly been considered as the near-equivalent of Poussin's classicism in painting, although the two men probably never met and the former spent his whole life in France. The object of Mansart's architecture is not surprise but visual and intellectual satisfaction. The placing of each motive and the scale and proportions of each part are extremely inventive yet, once having been invented, seem not only right but inevitable. As in Poussin's paintings, everything is done according to

reason. In the same way the quality of the style depends on the tension set up between the imaginative power of the initial idea and the strictness of its means of expression.

The turning point for French architecture came with Bernini's visit to Paris in 1665 to discuss the design for the east front of the Louvre. This was a moment of truth for European art. It heralded the transfer of the artistic leadership of the western world from Rome to Paris. The greatest living architect and sculptor was invited by the King who was now Europe's most powerful ruler, to fulfil the most important civil commission of the day. Yet Bernini's plans were rejected, like those of the other Italian architects invited to submit. His arrogance irritated the French and he returned to Rome in a huff leaving only the portrait bust of the King as a tangible memorial of his visit.

The King's minister, Colbert, felt that French artists were now capable of standing on their own feet; the actual,

57. **Inigo Jones.** *The Queen's House, Greenwich.* 1616–18, 1630–35. This chaste and intimate house, built as an adjunct to the rambling Tudor palace which then lay between it and the River Thames, was the first English building designed in the classical style. It was originally two rectangular blocks joined by a bridge which spanned a public road running through Greenwich Park. The sides were filled in by Jones's pupil, John Webb, in the early 1660s. The colonnades either side were added in the 19th century, following the line of the former road.

56. (left). **Benjamin West.** *The Death of Wolfe.* 1770. Oil on canvas. 59½ × 84 in. (151 × 213 cm.). National Gallery of Canada, Ottawa. This picture, representing General Wolfe's death during the capture of Quebec in 1759, is generally taken to be the first history painting in modern dress. Such was no doubt its ultimate significance and it was certainly popular when it was exhibited at the Royal Academy in 1771, but West's original intention may rather have been to raise the traditional modern-dress battle-piece, hitherto considered a rather low genre, to the level of history painting. He composed the picture on classical lines, with borrowings from the Old Masters and by reconstructing of the scene not 'as it was' but 'as it ought to have been'.

58. **François Mansart.** *The Orléans Wing of the Château of Blois.* Designed 1635. The wing added by Mansart to the Château of Blois for Louis XIII's brother, Gaston d'Orléans, is elegantly precise in detail yet monumental in its total effect. Its ordered harmony epitomises the spirit of 17th-century French classical architecture.

59. **Jules-Hardouin Mansart.** *The Church of the Invalides, Paris.* 1680–91. S. Marcello, the Invalides and St Paul's Cathedral represent three different types of Baroque-Classicism current in the late 17th century. The baroque element in the Invalides probably sprang as much from a desire for magnificence and grandeur of scale as from any wilful defiance of classical rules. Of the three buildings, the Invalides undoubtedly carries the most authority with its superb open site and the vista leading up to it from across the Seine.

60. **Christopher Wren.** *St Paul's Cathedral, London.* 1675–1711. (Modern drawing to show a view of the Cathedral now obscured by buildings.) The contrast between the classical dome and the baroque west towers of St Paul's is almost gauche, yet the total effect is pleasing and the dome itself is one of the noblest Northern Europe, while the quality of the detail in the west towers has been revealed by the recent cleaning.

91 **88** extremely classical east front of the Louvre was designed by Le Vau and Claude Perrault in collaboration; and when it came to finding an architect for the first re-building of Versailles for Louis XIV in 1668, there was no question of appointing anyone but a Frenchman. Le Vau's project was grand, effective and classically organised in terms of simple masses with a straight skyline broken only by statues.

Not only the statues, however, but the whole spirit animating Versailles, which was principally intended as a sumptuous setting for the Court, implied a leaning towards the Baroque and, from the 1670s onwards, baroque grandeur and pomp applied to basically classical architectural forms began to predominate in France. The leading architect of this phase was Jules-Hardouin Mansart, the great-nephew of François, who began to give Versailles its present, overwhelming form in 1678 and created *59* the baroque-classical church of the Invalides from 1680–91. The latter stands in an open setting at the back of the wide, comparatively low buildings of the hospital. With its rich decoration, tall lantern and needle-like spike at the top, the dome of the Invalides stands out in baroque splendour when seen, as it can be, from almost any point in Paris.

The other country with a significant though less deep-rooted classical tradition in architecture was England. To begin with, England was artistically so backward that the handful of buildings put up in the first half of the century in the Italian Renaissance style, designed by Inigo Jones, shone out like good classical deeds in an otherwise naughty late medieval world. Yet, by a curious combination of chance and retrospectively-felt influence, the Queen's *57* House at Greenwich, begun by Jones in 1618, became in the broadest sense the model for practically all English

domestic architecture for the next two hundred years. It is almost a reproduction of a Palladian villa, but not quite. Its proportions are longer and lower and it is at once more reticent and more picturesque in feeling (the latter quality is admittedly as much due to the setting as to the building itself). Compared with Mansart's Orléans Wing at Blois, it is simple almost to the point of insignificance, and is conceived in terms of line and surface rather than mass. Despite its plan, it is not visually a cubic building but a series of four separate façades.

From about 1660–1715, however, a new tendency was superimposed on this type of Renaissance classicism. The demand in England, as all over Europe, was for grander and more official-looking buildings, including churches. The dominant architect of this phase was Sir Christopher Wren, who not only rebuilt St Paul's Cathedral and fifty-one City Churches after the Great Fire of London of 1666, but also worked on several great public buildings (Chelsea and Greenwich Hospitals, Hampton Court, etc.) and the Universities of Oxford and Cambridge. He knew more of contemporary European architecture than Jones, had better opportunities and was a more inventive planner. In the City Churches he applied classical principles and **86** motives to a basically Dutch, box-like type of church design with great ingenuity and variety. His project for St Paul's began as a Greek-cross (equal-armed) structure with a dome and portico in the Renaissance manner, was then adapted for practical and liturgical reasons to incorporate a nave, and finally assumed a semi-baroque form. The juxta- *60* position of the plain but massive classical dome with the baroque west towers is inconsistent but effective. A harsher, more idiosyncratic type of English Baroque can be seen in the buildings of Wren's pupil, Hawksmoor, and on a colossal scale, in Blenheim Palace by Vanbrugh. At Blenheim, huge wings joined by colonnades open out one after another from the central *corps-de-logis*, striding over the Oxfordshire countryside with the military genius and arrogance of the owner, who never lived to see it complete— the great Duke of Marlborough. It is Versailles in a colder, more practical and more Augustan climate.

Realism

61. **Gregorio Fernandez.** *Pietà.* 1617. Painted wood. Museo Provincial, Valladolid. Spanish polychrome sculpture had its roots in popular religious art and was only touched in passing by the great stylistic movements emanating from Italy and Flanders in the Renaissance and 17th century. Its appeal lay in its surface realism and direct emotional intensity, although Fernandez here shows an acquaintance with current fashionable styles in the gesture and expression of the Virgin.

THE RELATIONSHIP BETWEEN REALISM AND IMITATION

Realism has been a recurring theme in European art from late classical times almost down to the present. But when realism has been found to be as constricting as a previous, anti-naturalistic style had seemed artificial and distorted, then artists have again moved away from realism and towards fantasy, thus beginning the cycle over again.

It goes without saying that whenever realism has appeared it has taken its precise character from the style of the period in which it found itself. Indeed 'realism' has not always been the term used; sometimes 'naturalism' or —with relation to late Roman portrait sculpture—'verism' have been used instead. In the 17th century there were three main kinds of realism: the first associated with Spanish polychrome sculpture (i.e. sculpture in carved and painted wood); the second with Caravaggism in the broadest sense; and the third with Dutch painting.

With all three kinds the emphasis has lain on the exact and careful rendering of surface appearances. Thus realism is a term of narrower scope than imitation, which in 17th-century theory referred at least as much, if not more, to the truthful representation of *human* nature. It was no doubt chiefly in this sense that Poussin meant it when he described painting as 'an imitation by means of lines and colours on a flat surface of everything under the sun.' This all-embracing definition reflected the long classical tradition, going back to Aristotle, according to which all forms of art, even dancing, were modes of imitation. In that sense art was the enactment in formal terms of almost any human act, story or emotion. Realism, on the other hand, comprises less than this. It does not even necessarily imply the logical reconstruction of spatial relationships and proportions. Neither the first type of realism to be discussed—Spanish polychrome sculpture—nor even to some extent the second—Caravaggism—was entirely logical in this respect.

SPANISH POLYCHROME SCULPTURE

Spanish carved and coloured wooden sculpture was religious in subject-matter and in the 17th century was basically a popular art. Its motivating spirit was the fervid, superstitious piety of the Counter-Reformation in Spain (virtually no female nudes or mythological subjects occur in Spanish 17th-century art). Glass eyes, eyelashes of real hair, wigs, and even costumes of real fabric were widely used. The statues so made were destined for altars or to be carried in groups through the streets (as they are still in some places) during religious festivals. In the first half of the century, however, there were at least two sculptors, Gregorio Fernandez in Valladolid and Juan Montáñez in Seville, who extracted from this crude *genre* an art of extraordinary expressiveness and poignancy.

The statues they created for monasteries and local parish churches embodied an intensity of emotion, expressed in terms that were somehow both Gothic and baroque at once, not seen elsewhere in European art at this time. To the popular mind these statues represented Christ, his Mother and the saints in states of anguish or meditation as real, familiar, recognisable figures. As works of art, they are quite without the formal sophistication or idealisation typical of Italian and Flemish religious painting and sculpture of the period. They contain no references to the Antique and only slight traces of the conventions of pose and movement handed down from the Renaissance. The articulation of their bodies is not always correct and it is sometimes hard to see how the figures would stand on their feet.

Their appeal lies in two things: first, the directness with which the emotions are expressed, a directness which seems wholly unpremeditated and to be the result of intuitive sympathy with the feelings of the people for whom the statues were intended; second, the very great beauty of their execution. The sharpness and explicitness of the carving is almost medieval in character although the forms themselves have a roundness and fluidity which betray their true period.

The colouring is intense, variegated and to modern eyes often overdone. Bright pinks, greens, blues, browns and flesh tints are applied to the highly polished wood which gives the figures a gaudy look (admittedly the colour is often the result of later over-painting). Only the sincerity of the emotion staves off sentimentality. It sometimes does not even do that; these statues are, after all, the ancestors of the saccharine plaster figures sold in churches and religious bookshops today.

17, 18 The next time that carved and coloured wooden sculpture flourished was in Bavaria in the 18th century. Bavarian 18th-century sculpture has all the swaying refinement of the late Baroque and Rococo, a refinement carried to the edge of decadence by the greatest Bavarian sculptor of the period, Ignaz Günther, although his work, too, was designed for a popular audience.

CARAVAGGISM

Caravaggio has already been mentioned in this book from the point of view of the dramatic and emotional qualities of his art. However, he was regarded in the 17th century—first with approval, later with mounting hostility—as a realist. 'Everything in art is trifling that is not taken from life' was a favourite saying of his, according to contemporaries. But how far were Caravaggio's paintings realistic and to what extent were they done from life?

62 His early works show an extremely exact rendering, based on careful attention to textures and local colours, of objects he could have set up in his studio—arrangements of fruit and flowers, flasks of wine, furniture, clothes, swords and so on. He is even known to have owned a pair of bird's wings, which a fellow artist borrowed on one occasion. All these things appear in his paintings, especially in the early ones which are quite brightly coloured. It seems reasonable to assume that such pictures were painted in the way Caravaggio said they were, that is, with loving care and with at least the costumes and still-life objects in front of him.

However, the representation of figures from the life would have been more difficult, even when a model or models could be made to pose in the studio, and it would have been completely impossible to paint figures in violent action in this way. In fact, as 17th-century critics noticed, uncertainty in the treatment of figures in movement was one of Caravaggio's weaknesses. He could concentrate, as he did, on violent subject-matter involving outflung limbs and contorted faces, and could create a convincing image of *action* by this means; but this did not solve the problem of movement, which required an ability to relate figures in space to each other and depended on learning an acquired system of pictorial conventions. Since these conventions were despised by Caravaggio, his figures appear curiously still, almost frozen, despite their emphatic action. They also

62. **Michelangelo da Caravaggio.** *Rest on the Flight into Egypt*. *c*. 1596. Oil on canvas. 51 × 63 in. (130 × 160 cm.). Galleria Doria-Pamphili, Rome. This is an example of Caravaggio's early style, antedating the characteristic 'dark' manner he developed in his middle years, as a result of which he decisively altered the face of European art. His early paintings are executed in strong, clear colours almost without shadows and with loving attention to surface detail; see especially the wings of the angel. The pose of the angel, however, is still Mannerist.

63. **Francisco de Zurbaran.** *The Martyrdom of St Serapion*. 1628. Oil on canvas. 47½ × 40¾ in. (120.5 × 103.5 cm.). Wadsworth Atheneum, Hartford, Conn. This very remarkable work was painted for the Mercedarian Monastery at Seville, to hang in the room where the bodies of the monks were placed before burial. St Serapion was an English bishop martyred in North Africa in 1240. As a work of art, the picture combines the mood and lighting of Caravaggio with the tactile qualities of Spanish polychrome sculpture.

have virtually no realistic setting, only a wall of darkness
40 which comes down behind them. Judged as a realist, what
Caravaggio gives with one hand—truth of surface detail—
he takes away with the other—failure to put the details
together so as to form a convincing whole.

What he excels in is truth to the physical and psychological facts of a situation. Even in his late works, from which
picturesque costumes and still-life are almost absent, there
is an insistence on incidental details—the lacing of a
garment, the grain in a piece of wood, or the sharp edge of a
sword-blade—which is all the more telling on account of
the economy with which it is used. Indeed such details,
which correspond to the way the eye notices small things in
moments of crisis, are an aspect of Caravaggio's psychological realism which is at least as important as his exact but
limited depiction of appearances.

But he was a realist above all in his choice of cast. A
Magdalene or a St Catharine by Caravaggio is an ordinary
Roman girl dressed in (almost) contemporary clothes and
only recognisable as a saint by her attributes. More important still, Caravaggio abandoned the tradition, which had
prevailed throughout the whole of Italian art in the previous two centuries, of representing sacred figures as heroes.
At the risk of putting too modern a gloss on his work, one
could almost say that Caravaggio was the inventor of the
anti-hero in religious art. His Bible episodes take place in
dark corners, his Christs and Saints are dressed in drab
clothes and are often half-obscured by shadow. They are
not larger or purer than life but, on the contrary, tough
working men who would hardly stand out in a crowd.
Ordinary people press round them in defiance of the
Counter-Reformation doctrine that lay people could only
approach God through the intermediary of the clergy.
Moreover Caravaggio sometimes shows both saints and
peasants with dirty feet.

Realism of this sort was held to be disrespectful to religion
with the result that Caravaggio's altarpieces were regularly
rejected by the church authorities who had ordered them.
This was despite the fact that he received repeated com-
missions from churches as well as from aristocratic persons
with *avant-garde* tastes. It was also despite the fact that, as
we can now see, he was one of the great religious artists of
the 17th century.

Caravaggio's influence lasted longest in the Spanish
colony of Naples and in Spain. In Naples, his style was
turned into a tough, dramatic, heavily-charged loaded
manner, suggestive of prisons and dark caverns, that can be
called Caravaggesque-Baroque, the artist chiefly respon-
24 sible for this being the Spanish painter, Ribera, who spent
most of his working life in the city.

Caravaggio had himself visited Naples leaving several
pictures there, two or three of which may have been sent
on to Spain. If so, it is possible that Zurbaran and Velasquez
may have seen them, for without copies, at least, of
paintings by Caravaggio it is difficult to explain the early
styles of these two artists. Neither, however, was indebted

to Caravaggism alone. At the very beginning of the century
the *genre* painter, Juan Sanchez Cotán, had produced a **75**
number of remarkable still-life pictures in a highly realistic
style which had affinities with, but was quite independent
of Caravaggism. Moreover, the Spanish tradition of carved
wooden sculpture was at least as important as Caravaggism
for Zurbaran and Velasquez. The former's *Martyrdom of St* **63**
Serapion is a fascinating combination of Caravaggesque and
local Spanish realism. The motionless figure is silhouetted
in livid colours against a neutral dark background and its
surfaces are treated in extremely sharp focus so that it
almost gives an illusion of being a carved statue; you feel
that if you were to tap it with a finger-nail it would sound
like wood. At the same time the monumentality and simplicity of the forms, the face with its shadowed eyes, and the
harsh raking light, recall Caravaggio. It is a startling, in-
elegant, yet deeply moving image that could only be found
in Spanish art.

Velasquez's earliest works are even more deeply rooted
in the tradition of Spanish wooden sculpture, yet he began
quite quickly to evolve a broader, more sensuous style. He
produced some pictures which are Caravaggesque in
subject-matter as well as treatment, but right from the
beginning the marks of his personal genius can be seen.
This genius has two main aspects. The first might almost be
called a human rather than an artistic one, since it has to do
with his reaction to people. After his appointment as court
painter at Madrid in 1625, he spent most of his life producing portraits, mainly of the Royal family, both separately or in groups. As far as is known, Velasquez never
complained of his lot or criticised his patrons. He was a
court artist no less than van Dyck, yet the results in terms
of style and treatment were very different. Velasquez was
evidently a secretive and withdrawn man and his sitters as
he depicts them are equally uncommunicative. The pomp, **6, 24**
trappings and grandeur of majesty are all there, as is the
stiffness of Spanish etiquette, yet one always feels that the
royal persons are acting out their roles and are not quite
grand enough for the part. Their expressions are veiled and
suspicious; the little princesses, in the marvellous series of
portraits which Velasquez painted of them towards the end
of his life, try to assert their dignity, but are too genuinely
child-like to succeed. More than any other great artist of
the period Velasquez concentrates the expression in the
eyes and mouth; the rest of the face and the hands and body
are mute. Again and again there are hints of some personal
longing, of a desire for some genuine emotion to escape the
formal conventions of dress and bearing, but Velasquez
never for a moment lets us know what this emotion is. Even
his clowns and dwarfs leer and grimace, not spontaneously,
but in a way dictated by their deformed physique and the
role expected of them at court.

The famous *Las Meninas*, painted by Velasquez towards **64**
the end of his life, seems to sum up his position. The painter
stands proudly beside the canvas yet is scarcely on the same
social level as the maid-servants and dwarfs, let alone the

64. **Diego Velasquez.** *Las Meninas (The Maids of Honour)*. 1656. Oil on canvas. 125 × 108½ in. (318 × 276 cm.). Prado, Madrid. *Las Meninas* has the air of being one of the great puzzle pictures of the 17th century. A sitting for a portrait is in progress which has been interrupted: the Princess, who is the subject of the portrait (possibly with her maids, since it is a large canvas), turns with her attendants to face the King and Queen, who appear in the doorway and whose figures are reflected in the mirror. Such an intimate glimpse of royalty is very rare, yet the mood is highly formal and various ambiguities are set up in the relationship between the observer and the observed.

65. **Jan Vermeer.** *The Artist in his Studio.* *c.* 1665–70. Oil on canvas. 51 × 43¼ in. (130 × 110 cm.). Kunsthistorisches Museum, Vienna. This is another view 'behind the scenes' of 17th-century art, and it, too, withholds as much as it reveals. Vermeer is showing how he painted a 'Vermeer', i.e. bit by bit, since the head of the canvas is nearly complete while the rest of the surface is still blank. Yet he gives no clue as to the personality of the artist, or even whether the artist is meant to be himself.

princess. His place, rather, is with the nun and the priest, and his figure is overshadowed by the vast canvas which we see from the back on the left of the picture. There is visual as well as psychological ambiguity in this composition, which results in a highly original variation on the normal baroque relationship between the work of art and the spectator. Framed in a mirror on the back wall are the images of the king and queen who are standing in the position of the real spectator of the picture. In imagination the spectator has become the two royal figures before whom their daughter and members of the household, including the painter, are standing. For a royal portrait of the period it is a uniquely intimate scene. Psychologically, few 17th-century paintings show a more 'baroque' relationship between the imagined and the real world, yet the painting is not baroque in any other sense.

Las Meninas is unusual in another way, too, for it is a fascinating representation of a painter at work. Not that Velasquez reveals anything of his methods; on the contrary it is typical of him that he should conceal them. But the very use of the mirror and the choice of subject—that of another picture being painted—shows a new self-consciousness concerning the act of painting. This is the other aspect of the genius of Velasquez, for one feels that no other

66. **Jan van Goyen.** *Landscape.* 1626. Oil on panel. 12½ × 21¾ in. (31.7 × 55.2 cm.). Messrs Alfred Brod, London (1961). This painting illustrates the first phase in the development of realism in Dutch landscape. The elementary problems of proportion and coherence have been solved, but the treatment of light and shade is still schematic, the details are all painted according to the same rather stiff formula, and there is little suggestion of atmosphere.

17th-century artist was so original in his use of paint. For him paint only partly served illusionistic purposes; it was applied less to reveal and define forms, than slightly to veil them, and hence to attract attention to itself as paint. Examining the treatment of a richly brocaded dress, we feel that the shimmering brilliance of the surface is not just that of the dress itself, but of the paint apart from the dress. The brush-strokes are, as it were, a substitute for embroidered leaves and flowers, not an exact visual equivalent of them. Moreover Velasquez goes further than any other 17th-century painter except Rembrandt in subordinating the contours of forms to the patches of tone and colour which make up the picture surface even though his figures are often silhouetted in contrasting tones against a neutral-grey background.

To what extent can this be described as a realistic style? Despite its origins in Caravaggism it is clearly very different in its mature form from the style of Caravaggio himself. On the one hand, it is less illusionistic in the treatment of surface detail; on the other, it is optically truer in the treatment of the whole. From a psychological point of view it is much less dramatic, yet precisely because of its calmness and perhaps even its ambiguity, it is closer to everyday experience. Velasquez's style is not so much 'more' or 'less' realistic than Caravaggio's as realism of a different kind —a kind that was admired and taken up again in the second half of the 19th century.

DUTCH PAINTING

Dutch painting—the third type of realism—presents a simpler problem than the art of Velasquez. It embodied a realism of content in a new sense and presented for the first time a comprehensive visual record of the way of life of a whole people. Its accuracy may be less literal than is sometimes supposed, but at any rate it represents the picture that Dutchmen of the period wanted to see of themselves and their country. Almost all possible forms of *genre* paint-

67. **Pieter de Hooch.** *Courtyard of a Dutch House.* 1658. Oil on panel. 28¾ × 24½ in. (73 × 62 cm.). National Gallery, London. De Hooch was a more explicit painter than Vermeer. His brick-by-brick treatment of the surfaces of everyday objects epitomises the 'photographic' tendencies in Dutch Realism.

68. **Adriaen Brouwer.** *A Boor Asleep. c.* 1632–38. Oil on panel.
14 × 10½ in. (35.8 × 27 cm.). Wallace Collection, London.
Brouwer was a Flemish artist who spent the first half of his short
working life in Holland and the second half in Flanders, where
this picture was probably painted. He was the greatest master of
peasant *genre* in either country and was the true successor to
Pieter Bruegel. Humour, insight, irony and the familiarity that
breeds respect, allied to marvellous technical skill, are combined
in this study of a young peasant asleep.

ing in the widest sense of the word were practised; that is,
paintings that depict an environment. The Dutch countryside with its canals, dunes, farm-houses and windmills
constituted the subject-matter of landscape painting. The
Dutch coast provided the material for marine painting.
People from almost all classes, in farms, taverns, guard-
rooms and ordinary domestic houses, figure in *genre* painting. The produce and imports of Holland—flowers, fruit,
fish, game, cheese, carpets, glassware, and silverware—
made up the various kinds of still-life painting. Finally,
Dutch artists represented church interiors, streets and—in
some ways most important of all—people, for there was an
unprecedented demand for portraiture.

Leaving Rembrandt aside for the moment, the end towards which Dutch painting moved can be most con-
veniently called photographic realism. Each picture represents a section of the visible world. The plain, objective

treatment is consistent throughout, each part of the composition being painted with the same care under natural
conditions of light. The consequence of this was that the
colours, tones, proportions, and spatial intervals had all to
be studied from nature and carefully related to each other.

On the other hand, the methods of achieving this result
were not discovered all at once and even the most realistic
Dutch paintings depend on artifice. As one follows the
history of 17th-century Dutch painting one can observe
stylistic conventions succeeding one another, coming into
fashion and going out again, and all the time becoming
more flexible and less obtrusive. The first thing to achieve
was pictorial clarity. The easiest problem to solve in
realistic painting is fidelity to surface detail, so that the
earliest still-life pictures, for instance, show the separate **76**
objects with brilliant and literal accuracy, but without
much relationship between them. The first landscapes,
continuing to reflect the methods used in the previous
century, show a marked contrast between the minuteness
with which leaves and branches are rendered and the
artificiality of the composition.

The trend towards a new kind of realism began in the
decade 1610 to 1620, a decade dominated in portraiture by
Frans Hals. In contrast to the conventions of earlier periods, space in landscapes and *genre* scenes was reduced and
the viewpoint brought down to eye-level. Paintings of all
kinds also now began to contain fewer figures and objects
than before. In the 1620s, with artists such as van Goyen **66**
(landscape), Porcellis (marine painting), Brouwer (*genre*), **68,69**
Heda (still-life), colours were subdued and compositions **77**
unified by subtle relationships of tone. By these means it
became possible to extend space further into depth and
by the 1640s landscapes had acquired an openness and **64**
atmospheric calm that form a parallel with the ideal landscapes of Claude painted in Rome at the same time.

The degree of realism achieved at this stage was exact
but limited, since colours were subordinated to tone and
forms to atmosphere. The final stage in the progress of
realism was reached in the third quarter of the 17th century
—the 'Golden Age' of Dutch painting. This was a period
studded with well-known names: Ruisdael, Cuyp, Vermeer, de Hooch, Steen, Terborch, Van de Velde, and **69**
Kalf. All forms of painting had a new solidity and breadth.
The lessons regarding tone, atmosphere and space learned
in the previous phase were now supplemented by mastery
of colour and form—qualities that had hardly been seen
before except in portraiture and religious paintings.

Along with an almost unprecedented representational
accuracy in the rendering of the visible world went imaginative qualities that saved these paintings from being
dull. Ruisdael's landscapes with their great trees, rocks and **66**
heavy clouds sometimes have an almost baroque grandeur.
Cuyp favoured a warm golden light in contrast to the grey **74**
atmosphere preferred by other Dutch painters and more

(Continued on page 121)

57. Claude Lorrain. *Landscape with a Goatherd.* 1637. Oil on canvas. 20¼ × 16¼ in. (51.5 × 41.3 cm.). Reproduced by courtesy of the Trustees of the National Gallery, London. This is an early, comparatively naturalistic work by the subtlest of all painters of ideal landscape, Claude Lorrain. The treatment shows similarities with the artist's drawings from nature, but the elements in the painting are selected and composed in such a way as to create a feeling of enchantment and delight. This mood is enhanced by the cool silvery light, which irradiates the background and filters through the trees. The goatherd is an imaginary figure from a remote Arcadian past, making the picture an equivalent of the type of pastoral poetry which began with Theocritus and Virgil and was much written and read in the 16th and 17th centuries.

58. (opposite, above). **Annibale
Carracci.** *The Flight into Egypt. c.* 1603.
Oil on canvas. 48 × 90½ in. (122 × 230
cm.). Galleria Doria-Pamphili, Rome.
This is one of a set of six 'lunettes' painted
by Annibale Carracci and his assistants
for the chapel of the Palazzo Aldobrandini
al Corso, Rome. With its orderly
arrangement of carefully chosen idealised
forms—foreground trees balancing
middle-distance buildings, diagonal lines
in one direction answering different
diagonals in another, solid equalling void,
'nature' reciprocating 'art'—it can be
called the first true classical landscape
painting.

59. (opposite, below). **Adam Elsheimer.**
The Flight into Egypt. 1609. Oil on copper.
12¼ × 16¼ in. (31 × 14.5 cm.).
Reproduced by courtesy of the Trustees
of the National Gallery, London.
Elsheimer, a German artist working in
Rome, invented a new kind of poetic
landscape—small, intimate, jewel-like
and strongly evocative of a mood. He also
made a notable contribution to the
painting of moonlight. Elsheimer's art
was much admired and was widely
influential during the first half of the
17th century. Note that the original of
this picture is much smaller than the one
illustrated above it.

60. (above). **Peter Paul Rubens.** *The
Tournament. c.* 1635–40. Oil on canvas.
28¾ × 42½ in. (73 × 108 cm.). Louvre,
Paris. This is an example of the 17th-
century realistic landscape of the
Netherlands, given a baroque vitality of
style and a mysterious, romantic mood.
The dexterity of the brushwork is such
that forms are sometimes indicated by the
merest touches of paint on the surface, yet
everything is vivid, everything is firmly in
place. The work lies mid-way between a
sketch and a finished picture—a new type
invented by Rubens.

61. (above). **Claude Lorrain.** *Landscape with the Nymph Egeria* (detail). 1669. Oil on canvas. 61 × 78½ in. (155 × 199 cm.). Galleria Nazionale di Capodimonte, Naples. The picture was painted for Prince Lorenzo Onofrio Colonna and included in Claude's record of his compositions, the *Liber Veritatis*, as No. 175. The scenery is loosely based on that of Lake Nemi, near Rome. The painting shows the extremes of subtlety and poetry which Claude's style reached towards the end of his life. The softness of the light, the strangely elongated figures

and the delicacy of the transitions in both the colours and the tones embody a mood and style that are much more ideal than in the *Landscape with a Goatherd* (plate 57).

62. (opposite). **Nicolas Poussin.** *The Burial of Phocion* (detail). 1648. Oil on canvas. 46 × 70 in. (114 × 175 cm.). Collection: the Earl of Plymouth, Oakleigh Park, Ludlow. Painted for a Parisian merchant named Cérisier, this work, with its pair, the *Gathering of Phocion's Ashes*, refers to the story of an

Athenian statesman who was condemned to death, and his ashes ordered to be scattered outside the town, for his refusal to abandon an unpopular policy under pressure from the mob. The style contrasts with that of Claude's *Landscape with Egeria*; the sunlight is stronger, the forms are more sharply defined and regularly drawn, and nature is, as it were, given the lucidity and order of a mathematical proposition. It should not be forgotten, however, that the painting has the beauty of mathematics as well as its impersonality.

63. (left). **Julius Porcellis.** *Seascape.*
c. 1640. Oil on panel. 13 in. (33.5 cm.),
diameter. Museum Boymans van
Beuningen, Rotterdam.
64. (below). **Jan van Goyen.** *View of
Emmerich.* 1645. Oil on canvas. 26 × 37½
in. (66 × 95 cm.). Cleveland Museum of
Art, Ohio, (John L. Severance Collection).
65. (opposite, above). **Philips Koninck.**
View in Holland. c. 1670. Oil on canvas.
52¼ × 63⅛ in. (133 × 161 cm.). National
Gallery, London.
66. (opposite, below). **Jacob Ruisdael.**
Wooded Landscape. c. 1660. Oil on canvas.
41½ × 58½ in. (103 × 148 cm.). Worcester.
College, Oxford.

These four paintings represent various
types of 17th-century Dutch marine and
landscape painting. All are more realistic
than the landscapes illustrated as plates
57–62 and all represent Holland; they
may be described as prose, whereas the
others, inspired by the scenery of the
Roman Campagna, embody poetry of one
kind or another. But they are not artless,
nor are they free from pictorial conventions. The first two belong to the phase
known as 'tone painting', of roughly the
second quarter of the century; the colours
are restricted, compositions are simple
and the picture space is chiefly taken up
with sea and sky. The other two paintings
belong to a later stage. They are more
colourful and dramatic, and forms,
instead of being dissolved in atmosphere,
are given their full plastic value.

67. Jean-Honoré Fragonard. *Blind Man's Buff. c.* 1775. Oil on canvas. 84 × 77½ in. (213 × 197 cm.). National Gallery of Art, Washington. (Samuel H. Kress Collection). At one time in the collection of Fragonard's chief patron, the dilettante Abbé de Saint-Non, this painting belongs to the tradition of Watteau's *Le Bal* (plate 19) and ultimately of Rubens's *Garden of Love* (plate 47). However, the subject is more anecdotal and the treatment more decorative. The light atmosphere, picturesque setting (based on the gardens of the Villa d'Este at Tivoli) and easy, relaxed, composition, so diaphanous that a puff of wind might blow it away, are typical of French Rococo landscape painting.

68. **Thomas Gainsborough.** *Going to Market. c.* 1769. Oil on canvas. 48 × 58 in. (122 × 147 cm.). Reproduced by courtesy of the Royal Holloway College Council, Englefield Green, Surrey. With its asymmetrical composition and cursive, elegant lines, this is one of the most rococo of Gainsborough's landscapes. The mood is considerably less artificial than that of *Blind Man's Buff* and a moist, silvery haze, suggestive of the English climate, pervades the scene. The gentle rhythms of the composition recall Gray's *Elegy*: 'Along the cool sequestered vale of life They kept the noiseless tenor of their way.'

69. (over page). **Sebastiano and Marco Ricci.** *Monument to the Duke of Devonshire.* 1725. Oil on canvas. 85¾ × 64½ in. (218 × 164 cm.). Barber Institute of Fine Arts, Birmingham University. This painting, signed by both Sebastiano and Marco Ricci, is one of about two dozen similar works by various artists commissioned as a commercial venture by an Irishman, Owen McSwinney, living in Venice. He described the scheme in a pamphlet entitled: 'To the Ladies and Gentlemen of Taste...' The subjects were allegorical tombs surrounded by ruins and treated in a fanciful and flattering Late Baroque manner, in honour of various prominent Englishmen recently deceased.

70. (above). **Francesco Guardi.** *View of Venice with the Salute and the Dogana. c.* 1780. Oil on canvas. 13⅜ × 20⅛ in. (34 × 51 cm.). Reproduced by permission of the Trustees of the Wallace Collection, London.

71. (below). **Antonio Canale (Canaletto).** *Venice: the 'Bucintoro' returning to the Molo. c.* 1728–30. Oil on canvas. 30¼ × 49½ in. (77 × 126 cm.). Reproduced by Gracious Permission of Her Majesty the Queen, Windsor Castle.

Canaletto and Guardi were the successive masters of view painting in 18th-century Venice. Of the two, Canaletto is the more factual and precise, exposing all forms—buildings, bridges, boats, water, people—to the same clear, even light. Guardi is the more evocative and fanciful. In his paintings light tends to be uneven and to flicker capriciously over the surface. Both pictures illustrated here, taken from points of view almost opposite

each other, show famous buildings in Venice: Guardi (above), the Customs House, and the sumptuous baroque church of Sta Maria della Salute. Canaletto (below), the Campanile, St Mark's Library, the Piazzetta (with a glimpse of St Mark's Cathedral) and the Doge's Palace.

72. (below). **Richard Wilson.** *Snowdon seen from across Llyn Nantll.* 1776 (?). Oil on canvas. 39½ × 50 in. (100 × 127 cm.). Walker Art Gallery, Liverpool. Classical in its construction and rococo in its principal outlines, this landscape is one of the first mountainous landscapes of the 18th century and one of the first oil paintings to show a view in Wales. The sensitive rendering of light and atmosphere owes something to both Claude and Cuyp, but Wilson adds a new, pre-romantic note of reverie. As Ruskin put it: 'I believe that with the name of Richard Wilson, the history of sincere landscape art, founded on a meditative love of nature, begins for England.'

73. (opposite, above). **George Stubbs.** *Horses in a Landscape. c.* 1760–70. Oil on canvas. 39 × 74 in. (99 × 188 cm.). Ascott House, Wing, Bucks., National Trust. The air of static calm, and clear, sharp outlines of the horses are qualities similar to those found in Wilson's *View of Snowdon.* But Stubbs, the finest of all English painters of horses, was a more objective artist than Wilson. The frieze of horses seen against the simple, unfocussed landscape has a neo-classical simplicity comparable to the figures on a Wedgwood vase.

74. (opposite). **Albrecht Cuyp.** *Cattle and Figures. c.* 1655–60. Oil on panel. 15 × 20 in. (38 × 51 cm.). Reproduced by courtesy of the Trustees of the National Gallery, London. Cuyp was chiefly a landscape painter, but animals often appear in his pictures and here they dominate the composition. He fills the picture space with a warm golden light which spreads from a source just above the horizon. His forms are heavier and less elegant than those of the two 18th-century artists illustrated on this page, but both were probably influenced by him.

118

75. (below). **Sanchez Cotán.** *Quince, Cabbage, Melon and Cucumber.* c. 1602. Oil on canvas. 25¾ × 32 in. (65.5 × 81 cm.). From the Permanent Collection of the Fine Arts Society of San Diego, California. Painted at the very beginning of the century in Toledo, Spain, this is a picture of hallucinated, almost surrealist, intensity, apparently planned on a basis of pure geometry; the curve on which the pieces of fruit lie has been shown to be a parabola. The interest in still-life detail, especially the careful, item-by-item depiction of detached objects, is typical throughout Europe at the time, but no other artist could have arranged the objects to form such an unusual design or treated them in such an animistic way.

76. (opposite, above). **Floris van Dyck.** *Still Life.* 1613. Oil on canvas. 19½ × 30½ in. (49.5 × 77 cm.). Frans Hals Museum, Haarlem. This painting is similar to the previous one in that the objects as such are treated in meticulous detail, but their arrangement is more commonplace. The composition is also more crowded and domesticated; it is, as it were, a brightly-coloured inventory of food and household goods rather than a consciously thought-out design.

77. (opposite). **Willem Claesz Heda.** *Still Life with Goblet and Fish.* 1629. Oil on panel. 18 × 27¼ in. (46 × 69 cm.). Mauritshuis, The Hague. This picture shows still-life objects treated by the method used in Van Goyen's *View of Emmerich* (plate 64); that is, as in a 'tone painting'. The forms are carefully selected and self-consciously posed, colour is subdued and the design is balanced in terms of horizontals and verticals.

78. **Willem Kalf.** *Still Life with Jug and Fruit. c.* 1660. Oil on canvas. 28¼ × 24½ in. (72 × 62 cm.). Rijksmuseum, Amsterdam. This picture shows a change in Dutch still-life painting parallel with the change in landscapes during the same period (see plates 65, 66). The splendid baroque silver jug (a handsome example of 17th-century Dutch silverware) was clearly painted with as much pride in the craftsmanship which went into the object as self-congratulation on the pictorial skill lavished on its representation. The ripe lemons spilling out of the Cornucopia-like Delft bowl also show the exuberance of the Baroque, as does the build-up of the composition.

69 typical of the local climate. Van de Velde exploited the decorative possibilities of 17th-century ships by silhouetting their sails and masts against the sky and dwelling on their ornamental hulls with loving care. Portraiture, other than Rembrandt's (about which one can hardly generalise), was marked by a new exuberance, reflecting both the skill *25* of painters and the self-confidence of sitters; with Hals, it became a field for the brilliant display of brushwork. The *67* domestic interiors of de Hooch, are characterised by subtle effects of perspective. Still-life painters, who had previously dwelt on the minimum of precisely painted, telling details, now delighted in the sensuous qualities of *78* objects; a still life by Kalf is almost as much a manifestation of the art of display as a baroque altar.

However, it was Vermeer who was the most imaginative of this group of artists even though his imagination was controlled by that most un-baroque quality, extreme reti- *51* cence. For Vermeer a figure in an interior surrounded by a few carefully chosen objects—a chair, a musical instrument, a map on the wall—was almost as much an opportunity for the painting of optical values as a king or princess was for Velasquez.

65 Vermeer's *Artist in his Studio* makes an interesting com- *64* parison with *Las Meninas* of Velasquez. Both are pictures which reveal the painter's self-consciousness about his art and its problems. Vermeer typically has his back to the spectator; his world is less ambiguously observed than Velasquez's, but is more enclosed. It is a world of still figures who scarcely communicate with each other and hardly ever touch. The beginnings of the picture on the easel in the *Artist in his Studio* would suggest that Vermeer painted his own pictures stroke by stroke, yet they have complete atmospheric and tonal unity. Like Velasquez, Vermeer never outlines forms, but paints them in relationships of tone. When examined closely, these tones can be seen to be very slightly simplified, suggesting that light and shade had an importance for the artist independent of their representational function, in the same way that paint had an independent importance for Velasquez. It is not an accident that both painters were admired by the Impressionists; indeed Vermeer, who was forgotten after his death, was not rediscovered until their time.

Finally Rembrandt, who is of all artists the most difficult to fit into any of the stylistic trends of the 17th century. The key words that come to mind when thinking of Rembrandt's art are spirituality and truth. His spirituality is very much that of the 17th century and his emphasis on emotional expression is characteristic of the Baroque, but with Rembrandt expression is not always evident on the face. Many of his figures are withdrawn and contemplative and what they contemplate is not outside themselves but within. This reflects a contrast between Catholicism and Protestantism rather than between Baroque and non-Baroque. What Rembrandt does is to show the isolation of the individual, *70* but in a religious and metaphysical context. It is the complete honesty with which he represents the human situation

69. **Willem van de Velde the Younger.** *A Dutch Man-of-War Saluting.* 1707. Oil on canvas. 65¾ × 91 in. (167.7 × 230.5 cm.). Wallace Collection, London. The ability to combine minute accuracy in the rendering of detail with faultless control of atmosphere and space was the ultimate achievement of Dutch realism. Van de Velde was also adept at making the most of the decorative beauty of sailing ships.

70. **Rembrandt van Rijn.** *The Three Crosses.* 1653. Etching (3rd state). 15¼ × 17¾ in. (38.7 × 45 cm.). This masterly etching by Rembrandt is hardly realistic in the visual sense. In its emotional power and dramatic use of light and shade it is closer to the Baroque. Yet it is without baroque idealisation or heroics and is governed by absolute fidelity to psychological and spiritual truth. In the 4th state of the etching, possibly executed some years later, Rembrandt radically altered the design, making it even more expressionistic.

71. **Jean-Baptiste Greuze.** *The Village Betrothal.* 1761. Oil on canvas. 36¼ × 46½ in. (92 × 118 cm.). Louvre, Paris. This picture was exhibited at the *Salon* of 1761, where it was greeted with rapturous enthusiasm by a public eager to pore over the details of its subject-matter. Bourgeois life began to acquire self-consciousness in the mid-18th century and to find its chroniclers in novelists and periodical essay writers. Greuze was its painter, and his work is important not only for its own sake but also for its revelation of a new anecdotal attitude to art.

that constitutes the chief sense in which he may be called a realistic artist. Even on a lower level the accuracy with which he shows the effects of atmosphere on forms and textures is unsurpassed: if one carefully observes the head
27 and right hand of the *Old Man Seated*, one can see how the form is fully modelled; yet it is not so much the actual surface of the form which is depicted as the light and atmosphere through which that surface is seen.

Nevertheless, this observed external realism is only part of the inner spiritual realism with which Rembrandt depicts the predicament of man, and—no less important—
48 woman. The *Bathsheba* is at once a baroque, a classical and a realistic picture and, at the same time, more than all three. It is baroque in the luminosity of its modelling and chiaroscuro and in the hint of deep golden brocades in the background; classical on account of the dependence of the pose on an engraving after an ancient Roman relief; realistic in the rendering of flesh and texture and, still more, in its honest presentation of the complex psychological facts of Bathsheba's emotional dilemma as she contemplates King David's invitation. It is more than all three in that it sums up Rembrandt's concern for humanity.

REALISM IN 18TH-CENTURY PAINTING

Realism in 18th-century painting consisted chiefly in depicting bourgeois activity in a moral light. It was thus as much a matter of content and attitude as of observation and style. Its antecedents lay in 17th-century Dutch *genre* painting, which now became popular, especially in France, but it was more self-conscious than its sources and showed signs of claiming a new status in the hierarchy of categories. Theoretical argument was deployed on its behalf, Hogarth acting as his own publicist and Greuze finding a champion in Diderot, who also admired the 'honesty' of Chardin.

A further characteristic shared by these three artists was an interest in story-telling. Unlike Vermeer's or de Hooch's domestic scenes, Chardin's imply the moment before and after the one represented; looking at his pictures we feel that in a few minutes the house of cards will fall down, the meal before which grace is being said will be eaten, the child so lovingly dressed by the nurse will go out for a walk. 53 With Hogarth and Greuze this preoccupation with narrative became explicit.

Borrowing his method and his cast of characters from contemporary fiction and the stage, Hogarth developed a new type of picture-story, of which he was both author and 55 painter, illustrating what he called 'modern moral subjects'. Six or more episodes would be assembled, showing a rake's or harlot's 'progress' or the consequences of marrying for snobbery or money. In each case the *donnée* of the situation is some character defect or act of folly which leads inevitably to a disastrous end. Along the way there is much coarse humour, an indefatigable attention to detail and more than a touch of melodrama, but genuine feeling underlies the satirical approach.

Greuze, who owed something to Hogarth, was just as didactic in his intentions but far more solemn. Belonging to the second half of the 18th century rather than the first, he was affected by the sentimentality of his age. Encouraged by Diderot, he also attempted to inculcate morality by the direct method—that is, by showing virtue itself in action—not by the indirect weapon of ridicule employed by Hogarth. In a period which was beginning to concern itself with problems of education and social relationships and which laid a new stress on the importance of the family, the success of such paintings as *The Village Betrothal*, *The Dutiful* 71 *Son* and *The Dying Grandfather* was assured. Here was painting which the *Salon*-going public could pore over and discuss.

Considering Chardin, Hogarth and Greuze as painters, Chardin was incomparably the finest of the three. He possessed a sensitivity to texture and atmosphere, shown particularly in his still lifes, which lay mid-way between that of the Dutch painters and the Impressionists, while the figures in his narrative compositions are touched with Watteauesque reticence and grace. Hogarth, too, conceived his 'line of beauty' as a rococo S-curve, but the gestures and attitudes of his figures are sometimes stilted and his chiaroscuro and brushwork are late baroque. In his insistence on nature rather than works of art as the proper standard of excellence for the painter, and in his discussion of problems of representation, he was more advanced in his theory than he was in practice. As a painter Greuze was skilful rather than sympathetic, at least to modern taste. To convey his message he often depended on the vocabulary of expressions and gestures proper to history painting. By what has sometimes been unfairly regarded as poetic justice but probably reflected no more than a hardening of attitudes in the Academy, in 1769 he submitted a conventional history picture, but was still only accepted as a *genre* painter.

The Rococo

INTRODUCTION

The Rococo is the most attractive artistic movement between the Renaissance and Impressionism, yet partly for that reason it is the hardest to write about. Its charm is a *cliché* of modern criticism; its qualities are all on the side of fantasy, wit, sensibility and smiling ease. Unlike the baroque and neo-classical movements that came before and after it, it is amoral and intuitive, not didactic and intellectual. It requires no elaborate background of historical or theoretical knowledge to be enjoyed. Its essential object is to please.

A century ago the brothers Edmond and Jules de Goncourt evoked the flavour of French rococo painting in words which are hard to improve on but over-rate sentiment at the expense of style. For instance: 'Watteau renewed the quality of grace. It is no longer the grace of antiquity that we meet with in his art:...The grace of Watteau is grace itself. It is that indefinable touch that bestows upon women a charm, a coquetry, a beauty that is beyond mere physical beauty. It is that subtle thing that seems to be the smile of a contour, the soul of a form, the spiritual physiognomy of matter.' Or again: 'Voluptuousness is the essence of Boucher's ideal; the spirit of his art is compact of it. And even in his treatment of the conventional nudities of mythology, what a light and skilful hand is his! how fresh his imagination even when its theme is indecent! and how harmonious his gift of composition, naturally adapted, it might seem, to the arrangement of lovely bodies upon clouds rounded like the necks of swans!' These two artists have worn the labels hung on them by the Goncourts ever since; and who shall say that those labels are entirely wrong?

What needs to be re-emphasised, however, is the technical brilliance of these artists—in other words, the 18th century's discovery of style. A sense of style, even stylishness, is the distinctive quality of rococo art. It is a quality that shows itself in the mastery of a particular rhythmical movement, the irregular S-curve, cultivated for its own sake and handled with seemingly effortless ease. We are made aware all the time of a consciousness of artistry, a revelling in the beauty of the performance, a delight in 'the thing well done'. It was an attitude eminently characteristic of an age which brought the conduct of human relationships to the level of an art. It belonged with those new arts of the 18th century, arts which have to do with communication between two or at most a small group of people: conversation, letter writing, chamber music, dancing, manners, seduction. Thus the Rococo was neither a public art, like the Baroque, nor a solitary art, like Impressionism, but an art of polite society.

Although lacking a theory, the Rococo had a stylistic momentum of its own, making it almost impossible for artists who used its conventions to be wholly bad; the worst vices of rococo art spring from want of feeling rather than want of skill—at times it could be heartless. On the other hand, the same conventions, superbly adapted to

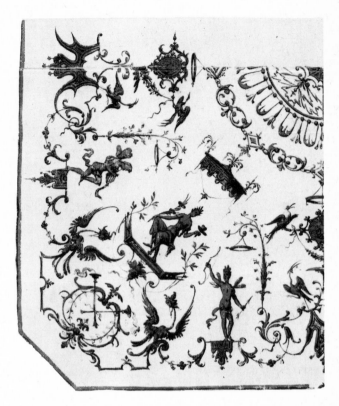

72. **Claude Audran.** *Ceiling Design* (detail). *c.* 1709. Watercolour and gold. 8 × 13½ in. (20.1 × 34.3 cm.). National Museum, Stockholm. This design for part of a ceiling decoration is similar to those produced by Audran for the Château de la Ménagerie (see text). The motives are adapted from arabesques (a traditional type of intricate pattern using a mixture of abstract forms interwoven with figures, animals and plants), but everything has been made lighter, freer and more playful. Such designs were an important source of the Rococo.

modest talents, put the very highest levels of imaginative achievement almost equally out of reach. Of the painters and sculptors of the period, perhaps only Watteau and Tiepolo can be compared with the greatest of their predecessors and successors—and Tiepolo was at least partly a late baroque rather than a rococo artist.

FRENCH ROCOCO DECORATION

It was in royal buildings dating from around 1700 near Versailles that the first examples of the new style of decoration were produced, and within two years it began to appear in certain rooms at Versailles itself. An important impulse behind this was the desire to escape from the stiffly formal atmosphere of the state apartments created at Versailles in the 1670s by Lebrun. As early as 1679 the first royal retreat, the Château de Marly (like so many other buildings of the period, since destroyed) was projected; for the next hundred years the search for ever more informal

73. *Salon of the Hôtel de Bourvallais (Place Vendôme), Paris. c.* 1717.
This is a typical French rococo interior of the Regency period
(1715–23). The hallmarks of this phase are the large mirrors
with curved tops, paintings over the doors, and repeated wall
panels decorated with stylised leaf motives at top and bottom
and elaborate medallions in the centre (the door panels are
similarly decorated). Notice also the S-curve of the fireplace
opening and the open-work cornice. As yet there is no break
with symmetry and the various units are still bordered with
straight lines. The colour scheme is white and gold. The architect
in charge was Robert de Cotte but the design of this room,
showing the influence of Pierre Lepautre, may be due to one of
his assistants.

settings continued, reflected in the construction of the
Grand Trianon, the *Petit Trianon* and the Belvedere, and
culminating just before the Revolution in the creation of
the English garden and the artificial rustic village in the
grounds of the *Petit Trianon* for Marie Antoinette.

A moment of significance for the genesis of the Rococo,
embodying an attractive personal touch, occurred in 1699.
In that year Jules-Hardouin Mansart, the chief architect at
Versailles and the newly appointed Surveyor of the Royal
Works, submitted a scheme for the decoration of one of the
rooms in the small Château de la Ménagerie, which had
recently been rebuilt for the thirteen-year old Duchess of
Burgundy, fiancée of the eldest grandson of Louis XIV. The
scheme provided for a ceiling painted with figures of
classical goddesses. The king, however, objected to these.
'It seems to me,' he wrote, 'that something ought to be
changed as the subjects are too serious; a youthful note
ought to appear in whatever is done. You will bring me

some drawings when you come, or at least some sketches.
There must be an air of childhood everywhere.' The
drawings, which survive though the Château does not,
were by Claude Audran, who later became the master of
Watteau. They show that the ceiling was covered with a
pattern of arabesques, more wayward than any previous
examples and revealing an almost calligraphic touch. The
motives consisted of loops and tendrils of acanthus leaves,
garlands of flowers, ribbons, arrows, stylised rustic bowers,
all spun out like filigree work, with hunting dogs, birds and
figures of young girls perched among the tendrils and
branches. The king was well pleased with the decorations,
which he called 'magnificent' and 'charming'. It should be
emphasised that the Rococo was welcomed by everyone
from Louis XIV downwards, eagerly and at once. It was
not until the 1740s that hostile criticism of it began.

Painted arabesques, derived ultimately from ancient
Roman 'grotesques' (so-called because they were found in

grottoes) and brought to a new pitch of refined elaboration in France under Lebrun, were a major source of rococo decorative motives. But the crucial step was the application of these motives to the carved frames of mirrors and wall panels. Here again the key date is 1699. The location was the redecorated king's apartment at Marly and the artist a draughtsman on the staff of the Office of Works, Pierre Lepautre. Hitherto the standard type of French interior decoration had consisted of wood or, less commonly, marble panelling divided into sections of rectangular or some other strictly geometrical shape. Above the panelling was a heavily moulded cornice and above that a high coved ceiling painted with allegorical figure subjects. Chimney-pieces were relatively high and the spaces above them usually filled with paintings or relief sculpture. Lepautre's innovations at Marly were, first, to modify the corners of the mirror-heads and panels by introducing leaf forms and curved scrolls in the shape of a C, both derived from arabesques; second, to alter the proportions of the mouldings so as to create a new effect of slenderness and grace. In the centre of the ceiling of the king's bedroom he placed a large rosette, leaving the rest of the ceiling blank—another idea which was to be widely imitated.

From this time onwards, mirrors—at first round-headed, later more elaborately curved at the top—became the standard feature above chimney-pieces. Chimney-pieces themselves were now reduced in height and their openings soon began to assume the characteristic S-curve of the Rococo. Architectural forms, where used (which was seldom), became increasingly attenuated. At the same time, doors and walls were raised and cornices pushed up into the ceiling so as to reduce the coving to a mere rounding of the angle with the wall. Later, mirrors were introduced not only above chimney-pieces but also on the other walls of the room apart from the window wall. Windows themselves, round- or oval-headed like the mirrors, were raised almost to the height of the cornice and were often continued down to the floor (hence the term 'french window'). Walls were flat, without projecting chimney-breasts or pilasters. Ceilings were left unpainted or were treated only with arabesques, while paintings with figure or animal subjects were confined to the spaces over the doors. The typical colour scheme was gold for the relief mouldings and ivory white for the walls. Sometimes the walls were left as natural wood, but positive colours were hardly ever used in France during the rococo period.

The rococo interior thus consisted essentially of a series of tall, slender, self-contained units placed side by side—mirrors, windows and doors alternating with decorative wall panels in carved wood. The technical term for such panels is *boiseries*. It was in the design of these that the rococo decorator (known as the *sculpteur*) showed his inventive skill. The corners of the frames were rounded and turned in, and the resulting C-shaped scrolls were made to issue in sprays of leaves pointing towards the centre of the panel. Mirror-heads were given increasingly complex curves with scrolls

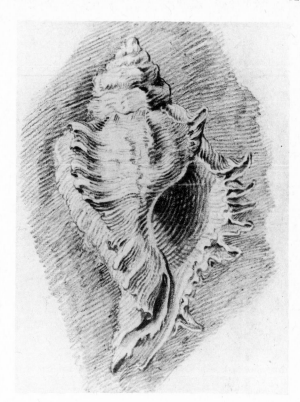

74. **Antoine Watteau.** *Drawing of a Shell.* c. 1710–15. Red and black chalk. $11\frac{5}{8} \times 7\frac{3}{8}$ in. (29.6 × 18.6 cm.). F. Lugt Collection, Institut Néerlandais, Paris. This drawing was probably made for documentary reasons, but the shell illustrated here may be taken as almost a microcosm of the Rococo. Many of the characteristic features of the style are shown—the irregular S-curve, the fronds and spikes, the asymmetry, the sensation of growth and the fascination with the natural world and with the exotic. Shell motives often occur in German rococo decoration, though not so much in French.

75. **Juste-Aurèle Meissonnier.** *Design for a Soup Tureen.* c. 1735. Engraving from the *Livre d'Ornemens.* $7\frac{1}{4} \times 10\frac{1}{4}$ in. (18.5 × 26 cm.). The intricate forms of this fantastic soup tureen, which would have been executed in silver, echo the natural forms of the shell drawn by Watteau. There is also a resemblance in spirit. Meissonnier's work is representative of the *genre pittoresque*, a phase of the Rococo characterised by a love of asymmetry and keen interest in the world of marine and plant life as a source of decorative motives.

76. **Germain Boffrand.** *Salon de la Princesse, Hôtel de Soubise, Paris. c.* 1735–40. This exquisite oval room is the *pièce de résistance* of the suite added by Boffrand to the order of the Prince de Soubise soon after the latter's marriage in 1732 to the 19-year old Marie-Sophie de Courcillon. The exterior architecture is of classic simplicity but the interior embodies the perfection of French Rococo. The ceiling of the room is pale blue, the walls are white and the mouldings are treated with gold leaf. Windows occupy four of the five arched spaces out of view of the camera in this illustration.

meeting back-to-back in the middle and further scrolls masking the transitions from the curved to the vertical parts of the frame. The centres of panels were decorated either with rosettes incorporating the same combination of scrolls and acanthus leaves, or with trophies, that is, assortments of instruments illustrating a particular theme or activity—music, geometry, astronomy, hunting, the visual arts, and so on. Although many of these decorative forms and motives used by the Rococo were of classical or Italian baroque origin, they were now employed in a new spirit and with a total rejection of the Antique. In contrast to the Baroque, where the forms sprang out of the mass, the Rococo relied upon a delicate surface play. Relief was everywhere kept low.

In 1701, two years after its first appearance at Marly and the Château de la Ménagerie, the new style was tentatively introduced at Versailles itself. This was in the apartments on the first floor facing the *Cour de Marbre*, which were remodelled at this time to form the *Salon de l'Œil de Bœuf*, the *Chambre du Roi* and the *Cabinet du Conseil*. Again the executive designer was Pierre Lepautre, and for the next ten or twelve years the Royal Works, of which he and the *sculpteur*, François-Antoine Vassé, were the leading spirits,

continued to be the main source of new ideas in rococo decoration. After 1715, however, when Louis XIV died and was succeeded by his five-year old great-grandson, Louis XV, the situation changed. The Regent, the Duc d'Orléans, transferred the hub of French society from Versailles to Paris. Private buildings—the Palais Royal (then owned by the Regent himself), the Hôtel de Toulouse, the Hôtel d'Assy, the Hôtel de Bourvallais, the Château de Chantilly and others—now became the settings in which the most advanced rococo decorations were executed.

It was during this period—the Regency, which lasted until 1723—that the stylistic features described above first became fully developed. The sophisticated, urban society which fostered them was that for which Watteau painted; its leading designer, in succession to Lepautre, was the Italian-trained Gilles-Marie Oppenord, whose father was Dutch though he himself was born in Paris. There was now a still greater use of curves, including contrasting curves, especially for the tops of doors and for the frames of the pictures above them. These frames took on an undulating, S-shaped outline, with palm branches for the mouldings. Shell motives were also sometimes introduced, although they never had the importance in French decoration that

77. **Jean Courtonne.** *Hôtel de Matignon, rue de Varennes, Paris.*
1722–23. French exterior architecture during the rococo period
was characterised by reticence and simplicity, although touches
of relief sculpture and channelled wall surfaces generally enliven
the façade. The 45-degree edge of the projecting centre-piece

seen here is also typical. The lack of monumentality is striking by
comparison with Blois by François Mansart (fig. 58), although
the tall windows fitted almost flush with the façade and the
short wings projecting at each end are elements in common.

they had in Germany. Altogether there was a new fluency
and ease, with more interlacing than before and an occasional touch, reflecting Oppenord's Italian training, of
baroque plasticity. Painted arabesques sometimes included
monkeys (from which the term *singeries* derives) and
Chinese figures. *Chinoiserie*, that typically 18th-century
flirtation with the exotic which accorded so well with the
Rococo, also became popular in England. Indeed in
England, where it invaded furniture design as well as
interior decoration, *chinoiserie* was one of the chief forms in
which the Rococo became acceptable (another was mock
Gothic).

On the Regent's death in 1723 the young king moved the
court back to Versailles, but the major initiatives in rococo
design were still taken in Paris. The late 1720s and 1730s
were characterised by the development of a new phase of
the style, the *genre pittoresque*, of which the principal feature
was asymmetry. The designer credited with the introduction of this was the ornamentalist, Juste-Aurèle Meissonnier, who was born in Turin of Provençal parents; his
opposite number as an interior decorator was the *sculpteur*,
Nicolas Pineau. It was Meissonnier, according to Cochin
in his attack on the Rococo in 1754, who 'invented the

contrast of ornaments, that is to say, he abolished symmetry', but he did this mainly in designs for craft objects in
silver, such as candlesticks, jugs, clocks, monstrances and
table ornaments. Engravings after his work were published
from the late 1720s onwards (this was a common practice
with rococo designers) and became widely influential in
Germany. They show an astonishing assortment of freely
curving, semi-organic forms, flowing, twisting and waving
almost like water plants, or like the tropical shell illustrated
on page 125 in a drawing attributed to Watteau.

Similar forms were used for the design of tables, chairs,
chimney-pieces, fire-dogs and clocks. It was an age in
which many new types of furniture were produced, each
decorated with marquetry or Chinese lacquer-work and
ormolu mounts (gold-leaf on bronze) often created by
famous craftsmen and designers—Cressent, Gaudreux,
Oeben, Caffieri and so on. The simple categories of table,
chair and cupboard used in previous periods were now
sub-divided into specialised types determined by their
purpose: writing tables, gaming tables, boudoir tables,
console tables, *secrétaires*, *bureaux* (both of these being types
of desk), armchairs, upright chairs, sofas, settees, ward-
robes, *commodes* (chests-of-drawers), bookcases, cabinets,

78. **Antoine Watteau.** *Sheet of Studies with two Figures and Hands. c.* 1715–20. Red and black chalk. 6½ × 8⅞ in. (16.5 × 22.5 cm.). British Museum, London. Appreciation of sketches for their own sake was a characteristic of 18th-century aesthetics and it is fitting that Watteau was one of the two greatest draughtsmen of the period (the other being Tiepolo). His drawings have all the delicacy and sensitivity of his paintings, but with an added charm due to their lack of finish. Despite their piecemeal character, the various studies of which the sheets are composed always belong together harmoniously on the page.

display cases and countless others, each with its own technical name. They reflected the new search for privacy and convenience, which occurred in society at this time.

By 1740 the climax of the Rococo had been reached in France, although the reaction against it scarcely began before the 1760s. One work belonging to this final phase must, however, be mentioned. This, perhaps the finest surviving rococo room in all France, was the oval *Salon de la Princesse* in the Hôtel de Soubise, a late work by Germain Boffrand, designed about 1735. Eight arch-topped spaces, either mirrors, windows or doors, are arranged symmetrically round the room with panels between their vertical sides and paintings by Natoire illustrating the story of Cupid and Psyche in the spaces above the panels. Above the whole scheme, already on the cove of the ceiling, runs a waving cornice which dips down to meet the tops of the arches and rises over the paintings to send out open bands of scroll-work which converge on a rosette in the centre of the ceiling. The room is more positively feminine than the private apartments at Versailles, as was appropriate for the young princess for whom it was created. From here it was a short step artistically to the still more ravishing and more flamboyant Hall of Mirrors by Cuvilliés at the Amalienburg, completed a few years later.

The unusual oval shape of the *Salon de la Princesse* is a reminder that the Rococo in France was essentially a style of surface decoration and not of three-dimensional architecture. Almost all the rooms so far mentioned in this chapter are similar in shape and proportion to those of the late 17th century, that is, nearly square in plan with the height about two-thirds of the length and width. The only important structural innovation was sometimes to round off the corners. Being a style without spatial qualities, it was not well adapted to the treatment of exteriors or churches.

Classical orders were sparingly used or omitted altogether and windows often had no pediments. The height of the house was reduced compared with the Baroque, with the main rooms often situated on the ground floor. The main form of exterior decoration was a light horizontal channelling of the walls and some judiciously placed relief sculpture. A new type of building, the 'retreat' or *maison de plaisance* in the park of a great palace, was also characteristic, and not only at Versailles. Examples range from the enchanting *Pavillon Français* in the grounds of the Trianon (c. 1750, by Gabriel), through the Amalienburg, to the comparatively large residence of Sans-Souci, built for Frederick the Great of Prussia by Knobelsdorff in 1745–51. All these are single-storey buildings with ground plans including projecting elements, either curved or jutting out in square forms at 45 degrees to the entrance front. In England they have their counterparts in Chinese, Indian and Gothic follies and sham ruins, though the latter were meant only to be looked at, not lived in, and were rococo only in spirit, not form. Similarly, the early 18th-century English landscape gardens of Pope and Kent, with their winding paths and streams, so different from the straight formal alleys of contemporary French gardens, may also be counted as manifestations of the Rococo, even though they surrounded houses of classical design.

FRENCH ROCOCO PAINTING

The development of French rococo decoration was due to a succession of comparatively minor figures, but French rococo painting was created by a single artist of genius, Antoine Watteau. Watteau's apprenticeship to the arabesque painter, Claude Audran, has already been mentioned; he was also linked with another painter of arabesques, Claude Gillot, who, in addition, was noted for his

79. *Mirror in the Chinese Taste. c.* 1740. Carved wood. Victoria and Albert Museum, London. This mirror, probably by an anonymous English craftsman, illustrates the fairly common fusion of rococo scroll-work and asymmetry with motives taken from Chinese art. The preference for natural over artificial or abstract forms is also typical of the period.

pictures of the Italian actors of the *Commedia dell'Arte*. Both these *genres* were important for Watteau. Like the arabesques of the interior decorators, his own work is intimate, curvilinear and graceful. Like them he also avoided the Antique except as a source of romantic, playful allusion. While he was growing up at Valenciennes (which until a few years before his birth was still part of Flanders), the disputes between the Ancients and the Moderns, the *Poussinistes* and the *Rubénistes*, the formalists and the colourists, were being decided in favour of the latter in each case, and by the time he arrived in Paris in 1702 Rubens had begun to displace Poussin as the hero of French painters. Watteau himself studied the great series of paintings which Rubens executed in 1621–25 for Marie de Médicis in the Luxembourg Palace, of which Audran was curator. His own style was in many ways a translation of Rubens's into **47** modern, lyrical terms, with the pomp and grandeur left out. But Watteau is not just Rubens on a small scale, nor is his work all grace and deportment and fashionable couples holding hands or dancing on a terrace or in a park. On the contrary, his art is intensely personal and rooted in reality, both the reality of visible nature and the psychological reality of human emotions and desires, independent of their religious and mythological sanctions.

Watteau's drawings are the records of his studies from **78** life. As a man he was restless and difficult and died relatively young of tuberculosis, condemned perhaps to be a spectator of, rather than a participant in the life he observed so sharply. He transformed that life into poetic terms without depending on baroque heroics. One feels that the transformation was conscious, even partly ironic. Herein lies the importance to Watteau of his interest in the theatre. His paintings seem to represent life changed into art in just the same way that life is changed into art on the stage, although his work is never stagey. (Hogarth made a similar use of the theatre, but with different results—results which *are* often stagey.) Significantly, Watteau's figures scarcely ever look at the spectator; they are joined in pairs and are absorbed in each other (but not self-absorbed like Velasquez's or Vermeer's). And when they turn to face us they reveal themselves as actors taking their bow, actors in the traditional roles of the Italian comedy: Pantaloon the merchant, Isabella and Orazio the star-crossed lovers, Harlequin the sprightly servant, Mezzetin the musician and, finally, Pierrot or Gilles, the dull-witted innocent whom Watteau raises almost to the status of a hero, perhaps the only hero in his art. The actual paint, which is handled with magical dexterity, further detaches the figures from everyday reality, making them untouchable and remote. Permeating everything is a sense of anticipation, of waiting on the event, of feelings never to be translated into action. There is another comment of the Goncourts', which beautifully sums up Watteau's attitude. Writing of the picture which is perhaps the masterpiece of rococo painting, the *Embarquement pour l'Ile de Cythère* (so-called, although this scene has been shown to represent a return from the island),

80. **Jean-Baptiste Lemoyne.** *Portrait Bust of Voltaire.* 1745. Marble. Life-size. Château de Chaalis (Oise). The weight of classical and baroque tradition was greater in sculpture than in painting or interior decoration, yet French mid-18th-century portrait busts have simple outlines and an air of relaxed civility that link them with painting. Lemoyne's bust of Voltaire has the sensuousness and surface vitality of the Baroque without its turmoil, though it is too bland to be a good characterisation. Voltaire was the presiding genius of the 18th century as Louis XIV was of the 17th, and the contrast in personalities is matched by the contrast between the two busts (this and pl. 22).

81. **Matheas Daniel Pöppelmann.** *Pavilion of the Zwinger, Dresden* (badly damaged in the Second World War but since restored). 1709–19. The Zwinger was an ensemble of buildings, pavilions and screen-walls surrounding a court; it was laid out by Augustus the Strong, King of Saxony, as an open-air pleasure palace for court balls, plays and other entertainments. The king's art collection and library were also to be housed there. What was built, though extensive enough, was only a small part of his intention. The type and sculptural character of the forms are strictly speaking baroque but the gaiety and lack of pomposity animating the scheme give it a rococo flavour.

they say: 'It is love, but it is poetic love, love which dreams and which thinks, modern love with its longings and its garland of melancholy.'.

Watteau died in 1721, leaving only weak successors to the *genre* he had invented, the *fête galante*. Perhaps his truest heirs were, on the one hand, Chardin, whose work has been mentioned in another context, and, on the other, Gainsborough, who may never have seen any of his paintings but who certainly knew his work through engravings and through the medium of drawings by French artists living in London. But Watteau was also a starting point for Boucher, the most successful painter of the mid-century and the favourite artist of Madame de Pompadour. Like Watteau, Boucher profited from the dethronement of the ancient gods which had been effected at the beginning of the century, but he continued to use them as protagonists of his pictures, not merely as benevolent statues half-hiding in the background, as Watteau had done. With Boucher the cloud of Olympus became the pillows and sheets of an

18th-century bed, and Venus and Diana were made into sex symbols of greater accessibility than ever before. The world of Venus in Boucher's imagination, with its sea-nymphs and shells, its coral, reeds, water and foam, was the perfect accompaniment to rococo schemes of decoration. The waves and gambolling nudes have the same sinuous outlines, while the cool blue of the sea, the white of the glistening spray, the applegreen of leaves and the pink and white of the bodies produce a typically rococo colour harmony. These same colours were also to be used in Sèvres porcelain, and Boucher's flesh-tints have a porcelain-like lustre. As a manipulator of compositions he attenuates the Baroque in ways that will now be familiar: the bodies group and intertwine themselves in a loose, asymmetrical yet continuous association, like the decorative motives in a rococo trophy, leaving a free irregular space around them. The *Cupid a Captive* illustrated here even has the tall vertical proportions of a rococo panel.

Boucher's even airier successor was Fragonard, who **67**

continued the rococo style in painting up to the Revolution, untouched by the neo-classical movement though sensitive to some early romantic tendencies. He belonged to the phase of *sensibilité* which swept across Europe, originally from England, in the second half of the 18th century. The titles of some of his pictures will indicate his characteristic vein: *The Stolen Kiss, Cupid stealing a Nightgown, The Souvenir*. He also loved games and picturesque anecdote. But together with Hubert Robert he rediscovered the Italian landscape for French art and treated it in a way that recalls Watteau rather than Boucher. Trees of soft, silvery green fountain up against the sky while boys and girls play in an Italianate garden below. This is rococo landscape painting at its prettiest and most assured.

On a more realistic level, the 18th century discovered the concept of individual personality (as distinct from temperament or character) and hence was a great age of portraiture. As might be expected, baroque trappings and psychological complexities were abandoned and in their absence the

individual was made to appear, alert and at ease, in his or her social aspect. There is no question of abandoning social status, but the sitter is now, as it were, prepared to converse with his social equals. Occupations—reading or embroidery—are sometimes admitted but they can easily be dropped if a visitor calls. Dress is simple and everyday, not that of the court; poses are relaxed and expressions animated. Nattier, Tocqué, Perronneau and Quentin de la Tour were the representative French portrait painters of the age, with John-Baptiste Lemoyne the Younger as the typical sculptor. In England their nearest equivalent was the Scottish painter Allan Ramsay. Perronneau and La Tour worked in a new, typically rococo medium: pastel (actually first used at the very beginning of the century). Pastel naturally lacks the density and luminosity of oil paint and would have been useless if the prevailing style had demanded the dark shadows of the Baroque. But the Rococo required instead that the artist concentrate on lightness of effect and use soft pale colours.

82. **François Cuvilliés.** *The Amalienburg, Schloss Nymphenburg, Munich.* 1734–39. The exterior of the Amalienburg suggests French elegance and simplicity, yet the rudimentary curved pediments over the windows would have been too whimsical for French taste, even at this period. The key word for the

building is 'charm'. The famous Hall of Mirrors (pl. 13) is in the centre, behind the projecting block. The Amalienburg was built for the Electress Maria Amalia, wife of the Elector Karl Albrecht of Bavaria.

THE ROCOCO IN GERMANY

The close-knit society of Paris and the French court produced a relatively homogeneous kind of art, at least in interior decoration. In the German-speaking lands (Germany itself, Austria and Bohemia) the situation was different. There the numerous kingdoms, principalities and bishoprics, spread over a wide area, made for great variety, and there were also other factors in operation to complicate the position. In the first place, many works were the product not of a single organising brain, as was usual in France, but of a team, which might well consist of artists of two or three different nationalities. The patron, too, would often take a hand in the design, which must sometimes have disconcerted the artists. Moreover, a 'baroque' situation still

obtained in many respects. That is to say, there was still a demand for prestige buildings on a colossal scale. Churches remained as important as palaces, the money for both often coming from the same source; and the baroque fusion of the arts, involving not only painting, sculpture and architecture but also, even more than in Italy or France, woodwork and stucco decoration, was of great significance. In fact, no French or Italian 17th-century work is quite so complete an example of the *Gesamtkunstwerk* ('total artwork') in the baroque sense as the interiors of Vierzehnheiligen or Ottobeuren or the *Kaisersaal* at Wurzburg, all of which date from the mid-18th century. Unlike the churches of Bernini and Borromini, which were not designed to be painted, almost all those by German and Central European

42

8,82

architects had painted ceilings—an idea derived from the late 17th-century decoration of Italian churches such as the Gesù and S. Ignazio. Indeed, 18th-century painting in Germany and Central Europe had little importance except in these decorative settings.

At the same time, the Baroque in these countries was not bound to the same intellectual traditions as it was in Italy and France, and was thus freer to develop in a rococo direction. 18th-century German interiors, whether religious or secular, became increasingly spacious and light. By a natural process of evolution the powerful, dynamic curves of the Baroque were able to assume the quicker, more playful and more sinuous rhythms of the Rococo, and rococo colour harmonies—white with pink, blue or green and relatively little gold—were adopted. Similarly, a baroque pulpit or altar-frame could grow rococo curls at its edges, while a pure rococo altar, with all that that implies of asymmetry, fantasy and caprice, could naturally form the climax of that architecturally most baroque of churches, the Vierzehnheiligen. It is not uncommon or surprising, either, to find rococo furnishings and stucco work in a baroque setting, as at Ottobeuren, or baroque decorative forms being used in a rococo spirit, as on the exterior of the Upper Belvedere in Vienna. An even better example of this is the Dresden Zwinger, which had no practical function whatever but was solely a place for entertainment and courtly display—the most expensive permanent playground in the world, until its virtual destruction in the last war. Still late baroque in form and dating from quite early in the century (1709–19), the architecture is hardly more than a skeleton on which to hang decoration that takes it close to the Rococo.

The tendency for decorative considerations to outweigh structural clarity can also be seen in the churches of the Asam Brothers in Bavaria, where the normal balance between the arts is tilted in favour of the sculpture, painting and ornament at the expense of the architecture, although here the spirit remains wholly baroque. Similarly in sculpture: Egid Quirin Asam's *Assumption* at Rohr acts as a link between the full Baroque of Bernini and the near-Rococo of Günther, which was itself a stage on the way to the pure Rococo of Nymphenburg porcelain. In this sequence one sees how the deeply serious, agonised ecstasy of Bernini is transformed first into the airborne, theatrical rhetoric of Asam, then into the plangent, affected yet personal grace of Günther, and finally into the pure icing-sugar sweetness of Bustelli, the designer of the Nymphenburg figures. At no point is there a definite break; there is evolution—in fact the spiritual and emotional distance travelled is enormous—but no revolution in style.

Sometimes, however, the movement of German Baroque towards the Rococo was combined with French influences, when it produced what in one sense was a mixed Franco-German style but which from another point of view was an extension of the French Rococo taken to extremes undreamt of in France itself. This development occurred in two main centres, Bavaria and Prussia, both of which were ruled by strongly Francophile princes. It was Bavaria that discovered the French Rococo first. As early as 1706 the Elector Max Emanuel, who had taken Louis XIV's side in the War of the Spanish Succession (and spent ten years in exile in Paris for his pains), sent the young Joseph Effner to Paris to be trained first as a garden designer and then, probably under Boffrand, as an architect. When patron and architect returned to Bavaria in 1715 they resumed the building of the palaces of Nymphenburg and Schleissheim (both begun earlier by previous architects), using the French manner and incorporating some restrained rococo motives in the interiors. More remarkable, however, were the results which followed from the replacement of Effner by Cuvilliés on the succession of Karl Albrecht to the Electorship in 1726. François de Cuvilliés, who was a dwarf, had been discovered by Max Emanuel in Flanders and sent to Paris to study from 1720–24, where his arrival coincided with the first full flowering of the French Rococo under the Regency. It was this style, not modified or tamed but developed in new and exciting ways, which he brought back to Bavaria, using it first in the *Reiche Zimmer* ('rich rooms') of the Munich Residenz, begun in 1729, then in the Amalienburg (1734–39) and finally in the Residenz Theatre (1751–53).

The hunting lodge and *maison de plaisance* of the Amalienburg, in the grounds of Schloss Nymphenburg, is Cuvilliés' masterpiece and by general consent one of the supreme examples of rococo architecture and decoration in Europe. Its centrepiece is a circular room, the Hall of Mirrors, about forty feet in diameter and with a domed ceiling. Adjoining this on either side are a bedroom and a gunroom which, with a kitchen and other service rooms, complete the building. Seen from the outside the wall of the central room projects in an undulating curve, while at the back the façade recedes concavely with further, tighter convex and concave curves either side of the door. The corners of the building are also cut out with concave curves. The decoration of the exterior is very simple. Graceful Ionic pilasters adorn the central features at front and back (otherwise the wall surfaces are partly channelled); simple 'bows'—the traces of pediments—are drawn like eyebrows over the windows on the main front, with sculptured heads in circular recesses between them; and the whole is finished off with a balustrade. Carefully placed rococo ornaments in relief sculpture complete the effect. It is all very lucid, elegant and almost entirely French.

Inside, however, Cuvilliés let his imagination take wing. The Hall of Mirrors is close enough in its general arrangement to the oval *Salon* in the Hôtel de Soubise to make one wonder whether they can have been designed independently of one another (though which came first is another question), but the differences are as striking as the similarities. Briefly, what Cuvilliés does is to take French rococo motives, adapted mainly from the *genre pittoresque*, and treat them with German plasticity and lack of inhibition. There

83. **Georg Wenceslaus von Knobelsdorff.** *The Music Room, Potsdam, Stadtpalais* (destroyed in the Second World War). 1745–51. Frederick the Great's passion for the French Rococo was almost as great as, if less enduring than, that of Max Emanuel of Bavaria. In the Music Room at Potsdam the compositional patterns of the *genre pittoresque* are combined with Chinese influence to produce an effect that is at once supremely graceful and romantically mysterious. Its pervasive flowing lines differentiate it from English *chinoiserie*, which was more self-consciously eccentric.

are no wall panels but a continuous sequence of oval-topped mirrors alternating, not quite symmetrically, with windows and doors. The bevelled edges of the panes of mirrorglass echo the glazing bars of the windows in a squared pattern which, enhanced by the reflections, creates an effect of all-round outdoor light. Everywhere the decorative forms are slightly heavier and a good deal more naturalistic than they would be in a French interior. Shells, lions' feet, flags, musical instruments, cornucopiae, plants and flowers make up the surrounds of the mirrors. As in the *Salon de la Princesse* an undulating cornice occupies the coving where the ceiling begins, but above this, instead of open bands of scroll-work leading to the centre, there is a fantastic display of luxuriant forms such as one might find in an overgrown garden—urns, baskets, *putti*, plants of all kinds, animals and birds, some of which fly off onto the ceiling. Cuvilliés maintains the rococo principle of keeping his ornament on the surface rather than making it grow out of the structure beneath, but when one looks at a room such as this one realises how clear, disciplined and restrained true French rococo decoration is by comparison. In the Amalienburg there are the qualities of both wit and elegance, but also a luxuriance, a readiness to go to extremes, that could only be found in Germany. Cuvilliés' decoration piles up in a firework display of ornament, all executed in silvered stucco on a pale blue ground. Beside it the equally personal and scarcely less beautiful decorations carried out a decade later by Knobelsdorff for Frederick the Great of Prussia look thin. In the interiors of Charlottenburg, the Potsdam *Stadtpalais* (partly destroyed in the last war) and Sans-Souci, there is a finer, lighter, more whimsical touch and still more asymmetry than in Cuvilliés' work— qualities which at Charlottenburg and Potsdam contrast with the classical architecture of the exteriors. The music
83 room at Potsdam shows Chinese influence; at Sans-Souci there were *singeries*.

Like Cuvilliés, Knobelsdorff had been trained in Paris and his interpretation of the Rococo is basically courtly and French, though adapted to German circumstances. The opposite combination—a native German Rococo tinged with French influences—is found in the work of the painter and plasterer, Johann Baptiste Zimmermann, who worked mainly in churches. After beginning his career as an assistant to Cuvilliés at the Munich Residenz and the Amalienburg, he went on to collaborate with his brother, the architect Dominikus Zimmermann. The typical Bavarian church interior of the mid-18th century—light, airy and resplendent with gay colours—was essentially their joint creation. These churches, of which the best known is Die
84 Wies (1745–54), were commissioned by local parishes and monasteries rather than courts and were spatially simpler than those of the Baroque. The architecture often seems cut on the curve like a gigantic piece of fretwork, which then becomes a surface for relief stucco and painting. At Wies, which is actually built of wood, the contours of the windows and arches are scrolled, the capitals are invaded by orna-

84. **Dominikus and Johann Baptiste Zimmermann.**
Pilgrimage Church of Die Wies, Bavaria. 1745–54. Architecture by Dominikus, stucco and painting by Johann Baptiste. Although the latter had worked under Cuvilliés at Munich and had thus absorbed French influences, a church like Die Wies can fairly be called German Rococo, that is, a local adaptation of the late Baroque. Scrolls are everywhere—on the arches, the windows surrounds, the capitals, the organ-case and, above all, the stucco *cartouches* (shields) round the pictures in the pendentives. Apart from their characteristic S-curves, these *cartouches* sprout into the fronds and spikes of a rococo shell (*see* fig. 74) and have the typical asymmetry of the *genre pittoresque*.

ment and rococo S-curves roam freely over the spandrels and vault, either as shell-like cartouches or as frames for the paintings. The colours are pink and green combined with gold and bluish white in the ornament, deeper blue and red in the columns of the choir and a dominant pale blue in the ceiling fresco.

The other important Bavarian church architect of the period was Johann Michael Fischer, who completed Otto- *82* beuren (in which Dominikus Zimmermann had also been involved) during the 1750s. The confused building history of this church did not allow for much architectural sophistication but it became the arena for some of the most dazzling rococo ornament in any German church. The stucco decoration and furnishings—altars, pulpits, statues, capi- *81* tals, picture frames and cartouches—are pinned to the wall surfaces like some vast collection of tropical butterflies. In the cartouches the rococo shell, with its irregular S-curves, asymmetry and growths of spikes and tendrils, found its ultimate decorative expression. The designer of all this was another specialist in stuccowork, even more important and brilliant than Johann Zimmermann—Johann Michael Feichtmayer.

PORCELAIN

Before leaving the subject of German Rococo, a word should be said about porcelain, whose manufacture in the mid-

18th century replaced that of tapestries as the prestige activity of courts. Until the early 18th century, 'hard-paste' porcelain, possessing the thinness and lustre of a shell and the hardness of steel, had been a prized Chinese import, the methods of its production a mystery to the West. But in 1709–10 the secret was discovered in Dresden ·and the famous factory of Meissen was set up under the patronage of the king of Saxony, Augustus the Strong. For the next forty years German factories dominated European production, but by the late 1740s rivals had sprung up in Austria, France, Italy and England, often for 'soft-paste' rather than 'hard-paste' ware; in the second half of the century the greatest centre was at Sèvres, near Paris. Unlike their Continental counterparts, the English factories at Chelsea, Bow, Derby, etc., were established by commercial enterprise, not by royal patronage. In the earliest examples at Meissen only 'useful' objects were made—plates, cups, saucers, jugs and bowls—in imitation of their Chinese originals, but soon figures began to appear as well. Sometimes these were designed by specialists such as Kändler and Bustelli, sometimes by artists who were primarily marble sculptors, like Falconet. Bustelli's Nymphenburg figures are in some ways the quintessence of the European Rococo, fusing all its expressions into a single miniature art. Stylistically their origin is not French but Italian, transmitted through the sculpture of Ignaz Günther, and their subject-matter is often taken, like Watteau's, from the characters of the Italian Comedy. In feeling they show the wit, grace and aristocratic elegance of the French Rococo.

THE ROCOCO IN ITALY

Owing to the Church's dominance over secular princes, the long classical traditions of Italian art and the overwhelming prestige of the great monuments of the past, which acted like a conscience in the minds of contemporary artists, the Rococo had little chance to develop in Italy, especially in Rome. Nor was it much helped by French influences, which tended if anything to reinforce the classical rather than the rococo trend. Yet, as in Germany, the Baroque—and even classicism—could sometimes mutate in a rococo direction or be combined with rococo elements. This was more likely to happen at a distance from Rome. Thus the late baroque churches of Juvarra and Vittone in Piedmont are painted in pale, varied, near-rococo colours, while the closest approximation to the Rococo in Italian sculpture was found at the opposite end of the Peninsular, in the work of the Sicilian sculptor, Serpotta. One may also include here the picturesque coloured figures made in Naples for Christmas cribs, which have something of the same origins and popular flavour as German polychrome sculpture. But rococo elements are not lacking even in Rome itself. The Spanish Steps are designed on a basis of lilting rococo curves, while the massive figure of Neptune on the Fontana di Trevi stands on a fluted scallop shell, and many anonymous 18th-century palaces have elegant rococo window frames connecting the different storeys

rather than elements of a classical order. In a house at Portici near Naples an entire room was decorated in the early 1750s with pure rococo ornamentation carried out in porcelain. Rococo rhythms and motives are also common in 18th-century Italian furniture design.

But it was, of course, in Venetian painting that the Italian Rococo found its greatest exponents. Venetian painting, which had sunk into near-oblivion in the 17th century, owed its 18th-century revival in the first place to the rediscovery of its own Renaissance past. Appropriately, its favourite artist from that past was Veronese. Tiepolo's style originally grew from a fusion of Italian late baroque influences and the figure types, colours and pageantry of Veronese. An early work by Tiepolo such as *The Trinity appearing to St Clement* has the rich colour, steep foreshortenings and powerful diagonals of the late Baroque, though the gain in lightness and decorative brilliance over the high Baroque of a 17th-century artist like Mattia Preti can be measured by comparing this painting with plate 44. When he reached Würzburg in 1750 to paint the ceilings of the staircase and the *Kaisersaal*, Tiepolo was the master of the most accomplished decorative style in Europe. Was that style rococo in the true sense? Hardly. Innocent of the naturalistic ornaments that made up the repertoire of Boucher's French Rococo, it kept more closely to the Italian tradition of dependence on the human figure. What is more, one feels that Tiepolo still sufficiently believed in the poetic and psychological truths of the classical allegories he was depicting—sufficiently, that is to say, for him not to need to sidestep them, as Watteau did, or, like Boucher and Fragonard, retain them only as a cover for erotic titillation. The old gods were still a living force and, in the Christian sense, Tiepolo was a great religious artist.

In terms of style Tiepolo used larger rhythms and broader brushstrokes than his French contemporaries and was still loyal to the traditions of the grand manner. But the admission that the greatest painter of the 18th century was technically behind the times need not worry us any more than the fact that the greatest musician of the period, Johann Sebastian Bach, was also out of date. The achievement of the *Kaisersaal* ceiling is breathtaking in its splendour and beauty. No other artist could have unified such a space with such consummate ease and grace; the sweeping, plangent curves that link the various groups appear to be simple and casual yet actually are subtle in the extreme.

(Continued on page 153)

79. **Carlo Maderna.** *Sta Susanna*, Rome. 1597–1603. The church of Sta Susanna, Maderna's first important work and his greatest artistic success, typifies Roman Early Baroque architecture. Although the general arrangement recalls Vignola's design for the Gesù, the width of the façade has been reduced in relation to its height, the various architectural elements (pilasters, half-columns, niches, volutes, etc.) have been made to fill more of the wall space and the design as a whole has been treated in higher relief. There is also a new concentration on the centre of the façade rather than its edges.

80. (left). **Francesco Borromini.**
S. Carlo alle Quattro Fontane, Rome.
1638–41. This little church, of which one
side altar and about a third of the interior
space are shown here, has been described
as one of the 'incunabula' of the Roman
High Baroque. Borromini based the plan
on an intricate geometrical diagram,
evolved through several stages, to which
he finally gave the form of an oval with
undulating sides (the two columns on the
right of the illustration mark the inward
projection of the curve). The dim light
and the height of the church in relation to
its small floor area give on the feeling of
standing in a shrine or grotto. The purity
of the architectural elements, unobscured
by sculptural decoration, is typical of
Borromini.

81. **Johann Michael Feichtmayr.**
Putto. c. 1760. Stucco. Abbey Church,
Ottobeuren, Bavaria. This bewitching
putto, perched on a curl of painted stucco
on the edge of an altar-frame, belongs to
the St John the Baptist altar attached to
the north-east pier of the crossing (on the
left facing the choir in the illustration
opposite). The religious statues in the
church are more serious than they look
from a distance, but they are surrounded
by decorative stucco-work of the utmost
fancy, ending in pure uninhibited play
at the extremities of each group. Feichtmayr worked at Ottobeuren from 1756
to 1764.

82. (opposite). **Simpert Krämer,
Johann Michael Fischer** and others.
Abbey Church, Ottobeuren, Bavaria.
1737–67. Like many Bavarian 18th-
century churches, Ottobeuren is by several
different designers. The somewhat conventional plan, with a nave, choir and
transepts meeting in a crossing under a
dome, is basically Krämer's. His work
was then modified by Fischer, who took
over in 1748 and was responsible for the
majestic, light-filled interior, in which the
spaces, articulated by regularly grouped
columns and a heavy cornice, flow into
one another in a typically late baroque
manner. With the decoration, however,
the idiom changes again, for, pinned to
the wall surfaces of the church, is some of
the most dazzling rococo sculptured
ornament to be found anywhere. This
ornament—playful, inventive and brilliant in white, gold and other colours—
was supervised and partly executed by
J. M. Feichtmayr.

83. (right). **François Mansart,** completed by Jacques Lemercier. *The Val-de-Grâce*, Paris. Begun 1645.

84. (below). **Francesco Borromini,** modified by Carlo Rainaldi and others. *Sta Agnese* in the Piazza Navona, Rome. Begun 1653.

These two church exteriors show a comparison and contrast between Roman and Parisian mid-17th-century architecture. The building histories of both churches are complicated. The Val-de-Grâce (right) shows obvious dependence on earlier Italian models (plate 79), but the lower storey (by Mansart), with its massive projecting portico, is more severe and classical than it would be in an Italian church. The dome also has an un-Italian solidity and compactness. Sta Agnese (below), commissioned by Pope Innocent X, is typical of the Roman High Baroque. The pierced towers of complex design at either end of the curved façade were an important and influential innovation. The effect, which depends on the tension set up between convex and concave curves and on the space 'held' between the towers and the dome, is brilliantly dynamic.

85. **Gianlorenzo Bernini.** *Tomb of Pope Alexander VII.* 1671–78. White and coloured marble. St Peter's, Rome. The Pope kneels calmly in prayer, allegorical figures wreathed in swirling draperies adore or mourn him, while Death, in the macabre figure of a skeleton with an hour-glass, emerges from the entrance of the tomb to summon him to eternity. As in the Cornaro Chapel (plate 1), the partly narrative, partly symbolic conception underlying the work is made vivid and actual by Bernini's inventive genius, backed up by immense illusionistic skill and the use of richly coloured marbles.

86. (above). **Sir Christopher Wren.**
St Stephen's, Walbrook, London. 1672–79.
This church, one of the finest of the
fifty-one built by Wren in the City of
London after the Great Fire of 1666, was
damaged in the last war and not all the
original gilding and carved woodwork
was replaced in the restoration. Apart
from its spatial complexity (a domed
central space combined with a cross-in-
square plan), the design has little in
common with Catholic churches of the

period. The effect is restrained, intimate
and dignified, but to these somewhat
pedestrian qualities the classically-
coffered dome and unbroken ring of the
cornice add a note of almost Renaissance
serenity.

87. (opposite). **James Gibbs.** *St-Martin's
in-the-Fields*, London. 1721–26. This view
shows the church as it was originally seen
before the construction of Trafalgar
Square, i.e. along a narrow street parallel

with the portico. The design is basically
derived from Wren's City churches but it
is in many ways a more assured artistic
design than any of them. (Gibbs was the
only English architect of his generation
with an Italian Baroque training). The
integration of the steeple with the church
so that it appears to rise through the roof
behind the portico was a daring, unclassical and ingenious innovation, which
became widely influential in North
America later in the century.

88. (above). **Louis Le Vau.** Château de Versailles: the Forecourt, *Cour royal* and *Cour de marbre*. Begun 1669. Although neither greatly distinguished as architecture nor fully baroque in style, Versailles —with its vast scale, magnificent interiors and gardens and, above all, its associations with the court of Louis XIV—is in many way the key secular building of the period covered by this book. The château on this, the town side, extends forward from a central block, the wings of which form first the *Cour de marbre* (originally built 1624 but altered by le Vau), then, second, opening out of the first, the wider *Cour royal* and finally the forecourt. The tall building on the right is the chapel added by J-H Mansart in 1689 to 1703—the most fully baroque part of the château.

89. (opposite). **Antoine Coysevox.** *Fame.*
1700–02. Marble. Jardin des Tuileries,
Paris. This allegorical equestrian
statue is one of a pair, the other
representing *Mercury*, made by Coysevox
for the gardens of the château of Marly, a
pleasure-retreat built for Louis XIV near
Versailles but since destroyed. As such it
may be taken to represent, at its best, the
garden sculpture ordered in immense
quantities for fountains, grottoes and other
ornamental purposes in the park of
Versailles. The turn of the figure's head

and the trophy beneath the horse recall
Bernini's equestrian statue of Louis XIV,
but the forms of the horse itself have been
regularised and given an almost relief-like
profile. This modification of baroque
energy and *élan* in accordance with
classical principles of style is typical of late
17th-century French art.

90. **Luigi Vanvitelli.** *The Great Cascade*,
Caserta, near Naples. *c.* 1770. Figures in
white marble. The palace of Caserta, the
'swan-song' of the Italian Baroque, was

built by Vanvitelli for the Bourbon King
of Naples, Charles III. In its scale and
impressiveness, its gardens, fountains and
cascades—though not in its plan or
style—it emulates Versailles. The
elegant pseudo-classical figures at the
bottom of the Great Cascade, representing
Diana and Actaeon, disport themselves on
the rocks as if taking part in a masque or
tableau-vivant. The group has no
architectural setting, and the baroque
union of art with nature here merges
into the Picturesque.

91. (above). **Louis Le Vau, Claude Perrault** and **Charles Lebrun.** *The East Front of the Louvre*, Paris. 1667–70. This severe, lucid façade, designed by three French architects, was chosen by Louis XIV for the east front of the Louvre after plans by Bernini had been rejected. Although the scale, the interplay of light and shade created by the free-standing colonnade and the variety of rhythm due to the coupling of the columns are baroque, the design is otherwise more strictly classical than that of any earlier French building.

92. (below). **Ange-Jacques Gabriel.** *Ministry of Marine*, Paris. Designed 1753, built 1757/8–68. The Ministry of Marine is one of a pair of buildings fronting the Place de la Concorde. Gabriel's basic plan, with its vista between the buildings leading to a projected domed Madeleine (never built in that form) was baroque. However, the design of the Ministry of Marine itself recalls the east front of the Louvre, while its greater richness, use of sculptural ornament and open arcades on the ground floor give it an 18th-century festive air.

93. (right). **Lukas von Hildebrandt** and **Johann Dientzenhofer.** *The Staircase,* Schloss Pommersfelden, Bavaria. 1714–19. Because by its nature it compels movement in three dimensions, the staircase as an architectural feature lent itself to baroque treatment, particularly to the use of contrasting intensities of light on its various levels. At Pommersfelden the staircase is the most prominent feature of the palace. The design was provided by Hildebrandt, its execution being entrusted to the resident architect, Johann Dientzenhofer.

94. (above). **Balthasar Neumann.** *Vierzehnheiligen*, Northern Bavaria. 1744–72. The plan and interior of Vierzehnheiligen ('Church of the Fourteen Saints'), by south Germany's greatest baroque architect, Balthasar Neumann, is of unparalleled ingenuity and complexity—a series of interpenetrating ovals and circles with no clear division between them and no dome over the intersection of the nave, transept and choir. The façade, of Gothic height and narrowness, and Borrominesque derivation in its architectural style, consists of two soaring towers flanking a curved front.

95. (right). **José Peña** and **Fernando de Casas.** *West Front of the Cathedral of Santiago de Compostela*, Galicia, Spain. 1667–1750. The baroque rebuilding of the exterior of the great romanesque cathedral of Santiago de Compostela, containing the shrine of the Apostle St James, was begun in 1650. Of the part seen here, the inner flight of steps leading up to the west door dates from 1606, the south tower, by Peña, from 1667–70 and the north tower (which repeats the south) and central section of the façade, which is by de Casas, from 1738–50. Architecturally, the most exciting part is the south tower, with its powerful silhouette and groups of five alternating square pilasters and half-columns, with unusual capitals on the storey below the belfry. The façade is typical of Spanish baroque architecture, a flat surface encrusted with a riot of fanciful decoration.

96. (above). **Versailles.** *Cabinet de la Pendule.* 1738. The *Cabinet de la Pendule* forms part of the 'Petits appartements' created in the 1730s by Louis XV in the north wing of Versailles facing the *Cour de marbre.* These rooms were deliberately planned to be lighter, pleasanter and more intimate than the 'Grands appartements' made for Louis XIV (plate 12) and are typical of French Rococo. The chimney-piece and furniture are of the period but the end of the room, originally oval, was squared off in 1760.

97, 98. **Joseph Willems** (?) *A Shepherdess* (97). $12\frac{1}{8}$ in. high (30.8 cm.). *An Actor in Turkish Costume* (98). $12\frac{3}{4}$ in. high (32.4 cm.). *c.* 1760–65. Chelsea porcelain. Victoria and Albert Museum, London. Meissen, the most important European centre of porcelain manufacture in the first half of the 18th century, prompted the growth of porcelain factories in England. Chelsea, the earliest of them, was apparently founded soon after 1745, its master-modeller being the Fleming, Joseph Willems. Willems's style is an elaborate interpretation of Rococo, loaded with picturesque detail. Shepherds and shepherdesses were the the stock themes of the late 50s and 60s.

99. **Robert Adam.** *The Etruscan Room*, Osterley Park, Middlesex. *c.* 1775. Adam claimed that 'A mode of Decoration has been here attempted, which differs from anything hitherto practised in Europe', i.e., the application of Greek vase motifs (which in the 18th century were thought to be Etruscan) to the decoration of an interior. In fact, the room has a rococo lightness and the forms and colours used bear a closer resemblance to Pompeian and Renaissance *grotesques* than to anything found in Greek pottery, but the archaeological spirit behind the design, stimulated by the discoveries at Herculaneum and Pompeii, shows one aspect of the neo-classical movement of the later 18th century.

100. **Richard Mique.** *The Temple of Love* in the Gardens of the Petit Trianon, Versailles. 1778. White marble. The picturesque 'Jardin anglais' laid out for Queen Marie-Antoinette in the grounds of the Petit Trianon was the final attempt by the French Royal Family to realise a century-long dream—the dream of escaping from the stultifying formality of Louis XIV's Versailles. The garden, with its winding paths, streams and lakes, contained—besides the Temple of Love—a grotto, a 'Belvedere' and a complete artificial village in rustic style (the 'hameau'), planned by the landscape painter, Hubert Robert, and executed by Mique. The Temple of Love itself, standing on a small island and glimpsed through trees, is reminiscent, like its English prototypes, of the buildings in Claude Lorrain's landscapes (plate 61).

85. **Francisco de Goya.** *The Parasol.* 1777. Oil on canvas. 41 × 60 in. (104 × 152 cm.). Prado, Madrid. This is one of the first series of designs for tapestries which Goya produced for the palace of El Pardo, Madrid, and with which he first made his reputation. The subject, the composition and the romantic, ironic charm recalling porcelain figures by Bustelli, are all in the spirit of Rococo. Yet a certain detachment in the artist's attitude to his subject-matter and a sketchy simplicity in his handling of paint hint at the great revolutionary painter he was to become.

He also adopted one technical device that was different from that used by 17th-century baroque ceiling painters; instead of making all forms converge on a single vanishing point, he used several vanishing points, which made the space more fluid but which at the same time were not noticed as an anachronism because of the lack of architecture and the large areas of open sky. Like Neumann's architecture and Bossi's stucco decoration, to which the ceiling forms a climax, this is late baroque form used in a rococo spirit, for pure enchantment and delight.

If 18th-century Venice lacked the repertoire of ornament of the French Rococo, it had its own equivalent—the caprice. 'Caprice' painting meant the assembling of imaginary ruins, statues, tombs, etc., together with monks, bandits and other characters then considered picturesque, in a fantastic setting. At the beginning of the century **69** Sebastiano and Marco Ricci produced a number of these caprices in a late baroque style, mainly for English consumption. The master of the pure rococo caprice—Francesco Guardi—belonged to the second half of the period. **70** In his views of Venice he gave the city a dream-like grace

very different from the bright, coloured-postcard sharpness of Canaletto, even though Venice, paradoxically, is one of **71** the few western European cities with no rococo architecture of its own.

In 1762 Tiepolo left Venice on his last journey abroad, to Madrid, where he died in 1770. His stay there partly coincided with that of Winckelmann's favourite painter, the neo-classicist Anton Raphael Mengs, whose followers executed the dreary canvases that replaced the great altarpieces designed by Tiepolo for the monastery of Aranjuez but taken down in the 1770s because their style was no longer fashionable. At the same time a young Spanish artist, Goya, was preparing to give the Rococo a last breath **85** of life in the tapestry cartoons he painted in the 1780s. The conjunction of these three artists, Tiepolo, Mengs and Goya—the first the master of the late Baroque and Rococo, the second the official representative of Neo-Classicism, the third the future genius of the Romantic Movement—in Madrid in the 1760s marked as crucial a turning point in the history of European art as Bernini's visit to Paris a hundred years before.

Neo-Classicism

A capsule history of art would state that the second half of the 18th century witnessed the rejection of the Baroque and Rococo and their replacement by a new style—Neo-Classicism—that depended on strict imitation of ancient Greek models, but this is so simplified as to be almost worthless; it not only misrepresents Neo-Classicism itself but also omits the Gothic Revival, Picturesque Movement and beginnings of Romanticism. It remains true, however, that Neo-Classicism is best defined in terms of its relationship to ancient art. It was a movement which embraced a new, more comprehensive and scholarly attitude to Antiquity than before, and which disregarded both the Middle Ages and the classicism of the Renaissance and the 17th century. This return to first principles is perhaps the central feature of Neo-Classicism. It implied that modern art was not just the latest manifestation of a long, continuous tradition but the outcome of a dialogue exclusively between the present and the distant past.

At least, this was the theory; in practice it took some time to realise and was perhaps never fully realised very often, least of all in painting. Other factors introduced further complications into the style, and it constantly overlapped and faded into the Romantic movement that grew up alongside it. Yet its ideal was to create a modern idiom based solely on research into, and reflection upon the principles of ancient art. This is more fundamental than the question whether Greek or Roman art was the main source. Looking at the problem in this light also reveals that the attitude to Antiquity was less uncritical than is sometimes supposed.

EARLIER 18TH-CENTURY CLASSICAL REVIVALS I: ENGLISH PALLADIANISM

An interest in the recent as well as the remote past is what distinguishes the classical revivals of different countries in the first half of the 18th century. In many respects these revivals were a continuation of the classical reactions of the 17th century. Their principles were the same: rejection of baroque (and where relevant rococo) extravagance and caprice, use of the 'best' models and a conscientious regard for rules. The first to occur in the 18th century began in England about 1715. This movement, usually called Palladianism, was almost wholly confined to exterior architecture, although it had some parallels in sculpture.

It opened typically with a manifesto, the first volume of a book, *Vitruvius Britannicus*, containing 100 engravings of 16th- and 17th-century buildings in Britain, chosen and edited by the Scottish-born architect, Colin Campbell. At the psychological climax of the book Campbell inserted illustrations of a house designed by himself, Wanstead in Essex, which was then nearing completion (it has since been destroyed). The implication was clear: although the earlier achievements of English architects were worthy of commemoration, the style of the future was to be that of Wanstead, with its clean lines, avoidance of baroque orna-

ment and visual drama, and straight façade dominated by a classically correct hexastyle (six-columned) portico. The words 'Vitruvius' and 'Britannicus' in the title of the book were both significant: the first recalled the first-century B.C. Roman writer whose *De Architectura* enshrined the rules of classical architecture; the second expressed the idea, widely held in England at the time, that the English were the successors of the Romans.

The 17th-century English architect regarded by Campbell as the closest follower of Vitruvius, and hence the best model from the recent past, was Inigo Jones; Wren, Vanbrugh and Hawksmoor were too baroque. Behind Jones lay Palladio, who gave the new movement its name. A new English translation of Palladio's *Quattro Libri dell' Architettura* also began to appear in 1715. The line, Vitruvius-Palladio-Inigo Jones, was to dominate English architecture, particularly country house architecture, for the next forty years.

Among subscribers to *Vitruvius Britannicus* was the young 3rd Earl of Burlington, who had just returned from the Grand Tour, already with a passion for architecture. In 1719 he re-visited Italy to study Palladio. While there he acquired a large collection of Palladio's drawings and later bought and financed the publication of drawings by Jones. He also invited William Kent, then studying in Rome to be a painter, to return to England to complete the decoration of Burlington House in London (now, much altered, the home of the Royal Academy). As a painter Kent was a failure but he became, in close association with Burlington, one of the principal architects of the English Palladian movement; he was also its leading interior decorator, furniture designer and landscape gardener. As such he was one of the first artists to take complete charge of the interior decoration and furnishing of a house.

Burlington himself was a considerable architect, and his own house at Chiswick, begun in 1725, illustrates the qualities of English Palladianism. Its model was a type of villa created by Palladio in north Italy, with a portico on the main front, rooms arranged round a central octagon, and the whole, except for the portico, fitted into a square. Its elevation is a freer interpretation of Palladio than its plan, but this in itself is characteristic. The qualities it exemplifies are: clarity, compactness and restraint; a liking for severe rectangular outlines, sharp angles and undecorated corners; an absence of columns or pilasters attached to the walls; insistence on plain lintels or unbroken triangular pediments over windows and doors; domination of the whole design by a classically proportioned and correctly detailed portico; finally, strict symmetry and harmonious grouping of the separate parts.

At Chiswick only the stairs, decorated with urns and rounded, sculptural balusters, relieve the austerity of the exterior with a touch of baroque movement. Yet this plain type of exterior architecture was often combined with a richly baroque treatment of the interior and an informal 'rococo' garden. This relationship between the exterior

86. **Colin Campbell.** *Design for Wanstead House, Essex.* Plate from *Vitruvius Britannicus*, Vol. 1 London, 1715. 9¾ × 14¾ in. (24.7 × 37.5 cm.). This is a plate from the book of engravings by the Scottish architect, Colin Campbell, which inaugurated the Palladian movement in English architecture. Wanstead, designed by Campbell himself and then in process of building, illustrates the style he advocated: lucid, regular, restrained in ornament, and correct according to the classical rules laid down by Vitruvius and followed (in his view) most faithfully in later times by Palladio and Inigo Jones. The house was demolished in the early 19th century.

design of the house and its setting was the exact opposite to that found (say) at Versailles, where a visually sumptuous architecture was accompanied by dead straight avenues of trees and rigidly geometrical *parterres*. Yet the solution at Chiswick was the more rational from the 18th-century point of view. To the generation of Burlington, Kent and Pope, who theorised about gardens as well as designing them, reason and nature were almost synonymous terms.

In their view the rational style for a building was necessarily the classical one, since a building is man-made, and it was self-evident that the controlled, classical approach to architecture was more in accordance with reason than the undisciplined baroque. In the 'natural' sphere of garden design, however, it was appropriate to apply as little control as possible, since nature was inherently reasonable and should therefore be left alone. Besides, the attractions of an unclipped and unstraightened-out garden were beginning to be felt. Although art was used, the object was to simulate a natural effect.

By his example and influence, which was political and social as well as artistic, Burlington established a near-dictatorship of taste in England until his death in 1753. Among contemporary architects James Gibbs, who had been trained in Rome under Carlo Fontana, was almost alone in having a strong enough artistic personality to withstand Burlington's influence—though even he modified the eloquent late baroque style he had learnt in Rome in a classical direction. St Martin's-in-the-Fields, designed by Gibbs in the 1720s, shows the successful results of this

process. With its stately portico, massive yet restrained treatment of the side walls and magnificent steeple rising directly above the roof, St Martin's is generally agreed to be one of the handsomest churches in the classical style in London. In many ways it represented the fulfilment of ideas conceived but never fully realised in Wren's City Churches. It also has the unusual distinction for a London church of being decorated inside with late baroque stucco ornament executed by Italian craftsmen. In the second half of the century it had a wide influence in North America.

EARLIER 18TH-CENTURY CLASSICAL REVIVALS II: FRANCE AND ITALY

No equivalent movement on the Continent exactly coincided with the beginnings of English Palladianism, although the theoretical writings of Cordemoy, which cut deeper than any of the rather superficial ideas of the Burlington circle, were published in Paris as early as 1706. The French architect who corresponds most nearly to Kent in the 1730s (Kent's masterpiece, Holkham Hall, was started in 1734), was Ange-Jacques Gabriel. Gabriel's career began in the 1730s at Versailles and Fontainebleau and he succeeded his father as chief architect to the king in 1742, but his most important works date from the 1750s. Stylistically he is very different from Kent as he depended on different sources; the parallel between them lay in their revival of classicism through its interpretations in the recent past. What Palladio and Jones were to Kent,

87. **Ange-Jacques Gabriel.** *The Petit Trianon, Versailles.*
1762–68. Begun for Louis XV's mistress, Madame de
Pompadour and finished for her successor, Madame du Barry,
this small retreat on the Versailles estate is a French parallel to
English Palladianism. It is square in plan with undecorated
corners, correct classical detail and simple proportions. Curves

and 45-degree angles are avoided throughout. Although its
designer, Gabriel, mainly used Italian Renaissance and
17th-century French classical sources, the building also has a
slight 'English' look, owing to its proportions and straight
balustrade.

François Mansart—a distant forbear—was to Gabriel;
significantly, neither Mansart nor Gabriel visited Italy.

In works like the Ecole Militaire (original project 1750)
and the Place de la Concorde (formerly Place Louis XV,
designed 1753), Gabriel sought to bring back the nobility
and grandeur which had been largely missing from French
architecture during the rococo period. The planning of the
Concorde, with its grandiose space and bold vistas leading
into and out of the square, represented a conscious return
to the age of Louis XIV although lack of money prevented
the project from being finished as Gabriel intended.

92 The Ministry of Marine and its companion building
91 which face the square echo the East Front of the Louvre,
only in richer, slightly more ornate form. Open arcades on
the ground floor, sculptured trophies on the skyline and
reliefs in the pediments at either end of the façades give
these buildings a festive air. Yet Gabriel has broken none of
the classical rules; everything is correctly proportioned;
symmetry, order and clarity are preserved. In the realm
of public buildings, this was a realisation of one of the key-
notes in French thinking about the arts in the 1750s—the
desire to return to 'the good taste of the previous century'.

In Rome during these years there was also a return to a
grander, more considered classicism than that of the late
Baroque-Classicism of the beginning of the century. 56
Curved façades and broken pediments were abolished and
sculptural ornament was reduced to a minimum. The
source of inspiration was the early 17th-century Baroque-
Classicism of Maderna, whose façade of St Peter's was the 4
model for that of St John Lateran, designed in the 1730s.

Developments in sculpture were less consistent and, at
least up to the mid-century, more closely linked to the
Baroque. But in England it is possible to find parallels with
Palladianism in the portrait-busts and tombs of the
Flemish-born sculptor, Michael Rysbrack, whose sources 88
were the Antique and the 17th-century sculptors, Duquesnoy and Coysevox. In France the struggle for classicism was
waged from 1732 onwards by Bouchardon, who returned
in that year from a long stay in Rome; the style he evolved
was a blend of the Antique and Girardon, with occasional
influences from the Renaissance.

There was also an attempt to revive history painting in
these years. In England it was Hogarth, rather surprisingly
and in a sense *faute-de-mieux*, who took up this task in prac-

88. **John Michael Rysbrack.** *Study for the Monument to Sir Isaac Newton in Westminster Abbey. c.* 1731. Terrocotta. 8 in. high. (20.3 cm.). Victoria and Albert Museum, London. This is a model for the principal figure on the tomb of Newton which was designed by the Flemish-born sculptor, Rysbrack, in collaboration with William Kent. The vigour of the execution and the strongly marked surface reveal Rysbrack's baroque inheritance, but the figure is composed on classical lines and has been given a noble Roman head and a grave authoritative expression. This is one of the finest pieces of sculpture produced in England in the 18th century.

tice (it did not lack supporters in theory). In France the role of impresario was assumed by Mme de Pompadour's brother, the Marquis de Marigny, who became Surveyor of the Royal Works on his return from Italy in 1751. In Rome, Pompeo Batoni, who is better known for his portraits, returned in his history paintings to the style of the 17th-century Bolognese classicists, Annibale Carracci and Domenichino. The Renaissance painter who was adopted as the ideal and patron saint of this movement was Raphael, whose reputation had been set above that of all others by the critic, Bellori, in the late 17th century. In France there was also a renewed interest in Raphael's great 17th-century successor, Poussin.

THE TURNING POINT OF THE 1750S

The position around 1750 may therefore be summed up as follows. In England, France and Rome—less so in other parts of Italy and hardly at all elsewhere in Europe as yet—there was growing hostility towards the Baroque and, where relevant, the Rococo. The Baroque was coming to be seen as extravagant, tortuous and unruly, the Rococo as degenerate and frivolous as well, since it was associated with a degenerate society. Attacks on the Baroque began in England in the years after 1715; those on the Rococo started in France in the 1740s, culminating in Cochin's articles in the *Mercure de France* in 1754. The style proposed as a replacement of the Baroque and Rococo was a sober, rational classicism. The Antique was to be the main source of this, but 17th-century classicism was admired and was often the direct model.

Developments on these lines occurred chiefly in architecture, but they also affected interior decoration, sculpture, and to a smaller extent, painting. Nor did the link with the 17th century end in practice with the 1750s. On the contrary, painters especially remained dependent on its aid, though they would not always have admitted it, and the career of Jacques-Louis David, who at first owed as much to Poussin as to the Antique, shows how difficult it was to achieve a 'pure' neo-classical style. The lack of extant examples of ancient Greek and Roman painting compared with the many works of sculpture and architecture available contributed to this situation.

Nevertheless, a number of other things happened in the 1750s which led to a radically new phase of the revival, and with these it is fair to say that Neo-Classicism began. Almost without exception they were not works of art but publications—reminders that Neo-Classicism was at heart an intellectual and philosophical movement, fostered by new ideas which only later, partially and not always successfully found expression in practice. In architecture the neo-classical period was *par excellence* an age of unrealised and unrealisable projects, though these might also be attributed to its romantic tendencies.

The key publications of the 'fifties included the first volumes of the Comte de Caylus's *Recueil d'antiquités* (1752) and of the official record of the archeological discoveries at Herculaneum, *Le Antichità di Ercolana* (1755); Piranesi's *Le Antichità Romane* (1756); Robert Wood's *Ruins of Palmyra* (1753); Laugier's *Essai sur l'architecture* (1753) and Winckelmann's *Reflections on the Imitation of Greek Works of Art in Painting and Sculpture* (1755). Also in the 1750s the Englishmen, James Stuart and Nicholas Revett, visited Greece, although the first volume of their great work, *The Antiquities of Athens*, did not appear until 1762 and was forestalled on the Continent by J. D. Le Roy's less well produced *Les Ruines des plus beaux monuments de la Grèce* (1758).

The significance of all this new archeological information is that for the first time it seriously extended the range of men's knowledge of the ancient world beyond that of the well-trodden ground of classical Rome. Now other periods and regions came into view: classical Greece, territories settled by the Romans overseas, ancient Egyptian civilisation, and pre-Roman civilisation in Italy itself. The character of this information and the wealth of new theoretical ideas generated at the same time are best considered under three main heads: one, the revival of Rome as a creative centre of the arts; two, the passion for Greece; three, the formulation of a new theory of expressive architecture.

THE REVIVAL OF ROME

It goes without saying that Rome had never ceased to be a centre of the arts since the High Renaissance. However, the decline in patronage which began before 1700 meant that few great works of art were produced during the first half of the 18th century and the number of famous Roman artists was very small. Even De Sanctis, the architect of the Spanish Steps, is scarcely remembered except by specialists. The Academy of St Luke and the Roman branch of the French Academy went into suspended animation. Few great foreign artists studied in Rome, preferring Paris or North Italy instead. Meanwhile students at the French Academy were notorious for neglecting the Antique. The keenest visitors were English, including connoisseurs, Grand Tourists and writers as well as artists; among them, as we have seen, were Burlington, Kent and Gibbs.

Yet it was during these years that Rome was being prepared for her new role as the great museum of Europe. Every day more pieces of classical sculpture were unearthed and added to those already familiar since the Renaissance. The Popes eagerly supported this activity.

89. **James 'Athenian' Stuart.** *The Monument of Lysicrates, Athens.* Plate from '*The Antiquities of Athens*', Vol. 1, London, 1762. 16¾ × 10½ in. (42.5 × 26.5 cm.). This engraving is taken from one of the key publications of Neo-Classicism. The discovery of ancient Greek architecture was a decisive factor in the movement, although its influence hardly became apparent in contemporary architecture until the 1790s. The refined style of this engraving is almost as significant as its subject-matter.

The Vatican collections were supreme, and in 1734 Clement XII opened Europe's first public museum of antiquities in the Palazzo dei Conservatori on the Capitol. Its main stock was naturally still Hellenistic, Greco-Roman and Roman work, and Roman copies of Greek originals, not authentic Greek work of what is now called the classical period (6th, 5th and 4th centuries B.C.). The significance of this for Neo-Classicism will shortly become clear.

The casual and nostalgic attitude to ancient Roman architecture of the first half of the 18th century is well summed up in the view paintings of Pannini. These represent the ruins themselves accurately enough but show them from picturesque angles and often in imaginary relationships to each other. What Pannini offered his patrons were theatrically handsome souvenirs, not records of hard archeological fact. Something very different was created by Giovanni Battista Piranesi, who was an architect by training and used the sharp, clear media of etching and engraving, not oil paint. But Piranesi was not merely a dry copyist; quite the reverse. His treatment of Roman ruins was highly subjective and dramatic and in some ways misleading. Using a low viewpoint, bold diagonals and strong tone contrasts, he surrounded his buildings with wiry foliage and small, bizarre figures reminiscent of Salvator Rosa or Sebastiano Ricci, and threw them in startling silhouette against the sky. The extravagance of Piranesi's imagination, revealed even more clearly in his etchings of fantastic prisons than in his views of ancient ruins, was far from the neo-classic ideal.

Yet his many publications, dating from the late 1740s to the 1770s, were more comprehensive than any produced before. In some he made useful contributions to research and, despite his play with scale, using small figures to exaggerate the size of buildings, he never tampered with the archeological correctness of his views. More important, his

90. **Giovanni Battista Piranesi.** *View of the Temple of the Sibyl at Tivoli. c.* 1755–60. Etching. $16\frac{3}{4} \times 25\frac{1}{8}$ in. (42.5 × 63.8 cm.). Piranesi's bold dramatic style lies at the opposite extreme from Stuart's delicate linear precision, yet both men—one the advocate of Greek, the other of Roman architecture—were scrupulous archaeologists and both were equally important figures in the neo-classical movement. Piranesi's copious etchings, which invested Roman remains with a new glamour and excitement, were especially popular in England.

91. **William Chambers.** *Somerset House, London.* Begun 1776. This illustration shows the central feature of the east side of the main court. Somerset House is a grave, noble, learned and competently planned building, rather than an inspired one. Its sources are found in Roman, Italian Renaissance and French 17th- and 18th-century architecture, but not in Greek, to which Chambers, as an academic and a traditionalist, was strongly opposed. He was, on the other hand, an admirer of Piranesi, and there is some similarity in feeling between his handling of classical forms and Piranesi's etchings. His role in English architecture was comparable to that of Reynolds in painting.

presentation of classical architecture coloured the attitude of his own and succeeding generations towards ancient Rome. Rome had threatened to become merely irrelevant or boring; now it was exciting again. Piranesi's personal contacts and his audience were wide, particularly among Englishmen. He was a friend of Robert Adam, an ally of Chambers, an inspirer of Soane and (in France) an important influence on the first true neo-classical building, the Panthéon in Paris, designed in 1755–56 by Soufflot.

The hankering after imaginative grandeur, sometimes turning to megalomania, which pervaded the 'romantic' wing of Neo-Classicism, started with Piranesi. To put it another way, it was partly due to him that a neo-classical building could be both forbiddingly severe and romantically evocative at once. The result was something quite different from the rhetoric of the Baroque; it was a quality residing in the massively hewn blocks of masonry themselves, in the sheer, cliff-like columns and walls and in the cavernous spaces under porticoes and vaults—a quality that was eloquent not of some human or divine message but of the internal drama of stone.

This contribution by Piranesi was more important than his spirited but ill-founded defence, conducted in the early 1760s, of the innate superiority of ancient Roman architecture as compared with Greek. Perhaps his principal role here was to keep the lines of demarcation between the two styles clear. The consequences of this can best be seen in England, where Piranesi's influence was stronger than anywhere else except Italy. Both Sir William Chambers and Robert Adam, who dominated English architecture from the late 1750s to the 1780s, took Piranesi's side against the partisans of Greece.

In fact Chambers, whose chief work was Somerset House in London, was in some ways still a Palladian, harking back to Kent and Inigo Jones as well as to ancient Rome. Adam was a much more original architect, more picturesque and less 'correct', but he too used only Roman sources—or what he thought were Roman sources. In fact there was a confusion here. In the controversy on behalf of the two opposing styles the chief argument in favour of the Greek was that Greek architecture existed first and that Roman architecture was derived from it—a point that the partisans of Rome could hardly contradict. The latter counterattacked, however, by claiming that both Greek and Roman architecture depended on Etruscan, which grew up on Italian soil and ante-dated both.

Little was known of Etruscan civilisation and nothing of its architecture but this was not considered an objection; it was enough to establish its theoretical *priority*. In 18th-century arguments that side would win which could show that its chosen style went back the furthest in time, reaching out to the dawn of civilisation and the very origins of man. This fascination with the primitive occurs, in a different context, in the architectural theories of Laugier, which will be discussed later.

Meanwhile, despite its irrelevance to the main problem,

92. **Jacques Soufflot.** *The Interior of the Panthéon, Paris.* Designed 1755–56. Originally built as the Church of Ste Geneviève, the Panthéon was secularised during the Revolution and made into a mausoleum for France's national heroes. It is planned on a Greek cross with equal arms, apart from an extention for the portico at the entrance, and has a dome over the crossing. Its austere grandeur echoes Piranesi's interpretation of Roman ruins, while its interior construction, using columns rather than arcaded walls or tiers, reflects the functionalist theories of the Abbé Laugier (see text). It is generally regarded as the first true neo-classical building.

the search for Etruscan works of art went on. Soon enough, in the course of excavations in south Italy and Sicily, some 'Etruscan' vases were found. These vases were in fact Greek but their Greek origin was not realised until later. They were eagerly sought after, notably by Sir William Hamilton, the British ambassador at Naples, who presented his collection to the British Museum in 1772. Josiah Wedgwood seized on them as the inspiration for his pottery, which he began to manufacture in the early 1760s at Burslem in Staffordshire, afterwards moving to a new site near Stoke-on-Trent which he called Etruria.

Finally, Robert Adam incorporated motives from these 'Etruscan' vases in his interior decorations. He combined them with figures, borders and slender, abstract architectural forms taken from the mural paintings found at Herculaneum, and also added motives from classical grotesques, to form a composite but highly original style. Yet, so great was the prestige of 'Etruscan' art, that he called the style by this name, and few Adam houses dating from the 1770s or later are without an Etruscan Room. By a curious irony one of the comparatively few cases of actual Greek influence on 18th-century art was thought to derive from something else.

Chambers, Adam and Soufflot (the architect of the Panthéon) all studied in Rome in the 1750s. They were the representatives of a new trend, whereby a visit to Rome became the rule rather than the exception for foreign artists. Other Englishmen there at some point during this decade included the painters, Joshua Reynolds, Richard Wilson and Gavin Hamilton (who stayed the rest of his life), and the sculptor, Joseph Wilton. Among French painters were Hubert Robert and Fragonard; Joseph Marie Vien, the teacher of David, had already arrived in Rome in the 1740s and returned to Paris in 1750. One important German painter was also present, Anton Raphael Mengs, who was another permanent resident except for periodic visits to Madrid. Almost the only important north European artist of this or the next generation who never visited Italy was Thomas Gainsborough.

However, not all those on this list became neo-classicists, as the inclusion of Hubert Robert and Fragonard makes clear. Wilson primarily discovered Claude Lorrain; Reynolds studied Raphael, Michelangelo, and, in Venice, Titian, besides the Antique. Both these English artists show the continuation of the approach to classicism through its Renaissance and 17th-century interpretations. What they achieved by this was a classic elegance and gravity, combined in Reynolds's case with a baroque chiaroscuro derived from Rembrandt and with the colour and breadth of handling of Titian.

In his *Discourses* Reynolds was if anything *opposed* to Neo-classicism except in sculpture; his approach was still basically that of Bellori and the 17th-century French academicians. Other painters who shared the same tendency were the American, Benjamin West, and Greuze, who paid a belated visit to Rome in 1768. Except for the modern dress on which its fame depends, West's *Death of Wolfe* was a pure exercise in traditional academic history painting, with poses taken from Poussin and Van Dyck and expressions from Lebrun.

Greuze used the same method in his *genre* pictures. Despite the everyday subject-matter, the figures obey the classical rules, putting the correct, revealing expressions on their faces, making emphatic gestures with their hands, moving their limbs parallel to the picture plane and grouping themselves so as to underline the narrative quality of the picture.

In fact it was not until a slightly later generation (which included West)—which reached Rome in the 1760s—that the development of a tentative neo-classical style in painting and sculpture became possible. The leaders of this group were Gavin Hamilton and Mengs from the previous generation, who were in Rome already. Besides West, who used the new style for his pictures of classical subjects, the group included the painters, Angelica Kauffmann and James Barry, and the sculptors, Joseph Nollekens, Houdon and Clodion. Of these only the sculptors were artists of the first rank; the painters were interesting rather than good. However, by this stage a new factor had arisen—Winckelmann's panegyric on behalf of ancient Greek sculpture.

THE PASSION FOR ANCIENT GREECE

The supremacy of Greek art over that of all other nations and periods had been admitted by various artists and critics long before Winckelmann's time. Among the earliest to do so was Nicolas Poussin, but at that time it was still a question of speculation as almost nothing was known of Greek art at first hand (curiously, little attention was paid to the one major work of Greek architecture available to them, the great Doric temple at Paestum, south of Naples). A further impulse to the doctrine of Greek superiority was given by the English philosopher, Shaftesbury, in the early 18th century. It was no doubt his influence that prompted Stuart and Revett to visit Athens. Once they and others had made the journey in the 1750s, the separate character of Greek as distinct from Roman architecture became plain and the battle on behalf of the two styles was joined.

Winckelmann, who had also read Shaftesbury, tackled the problem in the more difficult field of sculpture. His contribution was twofold. First, he asserted the supremacy of Greek art over all later art with a new clarity and force. The opening sentence of the *Reflections* reads: 'Good taste, which is spreading more and more throughout the world, was first formed under Greek skies.' His second point, because it was newer, was even more decisive, though it was at the same time subtly confusing for reasons that will be seen in a moment. What Winckelmann claimed to do was to define the special characteristics of Greek sculpture. Here there are two key passages. The first reads: 'The outstanding, universal characteristic of Greek works is thus a noble simplicity and calm grandeur in action and expression. As the bottom of the sea lies calm beneath a foaming surface, so the statues of the Greeks express nobility and restraint even in suffering.' The second passage paid tribute to 'the precision of contour, that characteristic distinction of the Ancients.' Interwoven with these brief but telling sentences was a restatement of the old academic doctrines concerning decorum, idealisation, obedience to the rules, and so on, taken from Bellori. It need hardly be added that the whole book was violently hostile to the Baroque.

Not the least remarkable thing about the *Reflections* is that they were written not in Rome, still less in Athens, but in Dresden, where there were only minor examples of classical sculpture and weak copies of the famous masterpieces in Italian collections to be seen. It was not until after the book was published that Winckelmann moved to Rome, to become librarian to the great collector, Cardinal Albani, and in 1763 Keeper of Antiquities at the Vatican. Yet even in Dresden he was able to see beneath the surface

93. **Etienne Falconet.** *Baigneuse.* c. 1760–80. Marble. 15 in. high (38 cm.). National Trust, Waddesdon Manor, Bucks. For much of his life Falconet was Director of Sculpture at the Sèvres porcelain manufactory, and his statues and figurines have the polished lustre and rococo sweetness combined with classically simple outlines characteristic of Sèvres. Falconet maintained that the warmth and softness of the human body were better rendered by his own contemporaries than by the Ancients.

of the feeble objects in front of him, to grasp something of the essence of classical Greek sculpture that lay within. 'Noble simplicity', 'calm grandeur', 'precision of contour' —reading these phrases nowadays one is apt to think automatically of the Parthenon frieze. But the sculptures from the Parthenon did not become well known to the west until the 19th century, and Winckelmann's descriptions were meant to apply to later, Hellenistic, works like the Laocoön, the Apollo Belvedere and the Medici Venus.

When he got to Rome he wrote a series of appreciations of these works which bore out his original theories. Herein lay both the strength and the weakness of Winckelmann's influence. On the one hand he provided a novel, arresting interpretation of long-familiar masterpieces of classical sculpture which was comparable in its freshness and visionary enthusiasm to Piranesi's re-interpretation of ancient Roman architecture. On the other hand, his claim to distinguish between authentic Greek art and its later derivatives was undermined by the scarcity of actual Greek works of the classical period. (It was perhaps typical of him that he never visited Greece.)

On the lowest level he simply encouraged a confusion of terms; people came to regard as 'Greek' what they had previously called 'antique'. In France, especially, many of the forms hailed as *à la grecque* were still Roman in origin, even though they were treated in a 'calmer', 'simpler' and more 'precise' way than before. In sculpture the result of Winckelmann's writings was to induce a glacial, polished look, with all individuality eliminated from the surface and a characterless stiffness of pose. The unGreek but typically mid-18th century warmth of sentiment which he unconsciously introduced into his descriptions also had surprising consequences. By an unexpected paradox it became possible to combine 'Greek' smoothness and precision of outline with rococo eroticism, most noticeably in female nudes: hence the 'pure' yet seductive girls in paintings by Vien with titles like *Greek Lady at the Bath*, and the coy nymphs and bacchantes of Falconet and Clodion.　93

In short, smoothness of form and clarity of outline became the distinctive qualities of neo-classical painting and sculpture, and fidelity to Greek or Hellenistic models was of secondary importance. A clear illustration of this is the 'demonstration piece' of the neo-classical movement in painting, Mengs's *Parnassus*, which was executed in 1761　94 for a ceiling in the Villa Albani, with Winckelmann's approval. Although it is a ceiling picture it is not treated illusionistically but as if it were to be hung on a wall. The principles on which it is painted are Winckelmann's but the 'imitation of the Ancients' which he urged as 'the only way for us to attain greatness' is played down. The background space is shallow, the movements are restrained and kept parallel to the surface, all the contours are distinct, not overlapping, and the composition is spread out like a classical frieze. But the sources used for the figures include not only the Apollo Belvedere and ancient Roman paintings from Herculaneum but also pictures by Raphael and Poussin. In fact the whole painting is more reminiscent of Raphael than anything else. The same reliance on a frieze-like composition and a mixture of sources is found in the first clumsy neo-classical history paintings of the 1760s by Hamilton and West. These pictures constituted a *genre* and a style that only reached maturity with David's masterpiece, the *Oath of the Horatii*, in 1784.

A more immediately successful application of the doctrine of linear precision can be seen in such things as engravings, interior decorations, works of small-scale decorative sculpture and objects of art. Some typical examples are Wedgwood vases, the early relief sculptures of Flaxman　97 (who designed for Wedgwood) and the beautiful friezes of horses in paintings by Stubbs, though the latter are **73** without direct neo-classical associations. More beautiful still are the supremely graceful interiors of Robert Adam,　95,9 where the quality of linear precision is combined with an ornate, 'all-over' treatment of the surface reminiscent of the Rococo. The decorative motives used in these interiors —Greek frets, Roman acanthus scrolls, anthemion and guilloche borders, sphinxes, griffons, rams' heads, masks,

94. **Anton Raphael Mengs.** *Parnassus with Apollo and the Muses.* 1761. Fresco. Villa Albani, Rome. Mengs's *Parnassus* is the most celebrated painting of the first phase of Neo-Classicism, although it is not the earliest and is not purely neo-classic. Its sources include the works of Poussin and Raphael as well as the Antique. The central figure, adapted from the *Apollo Belvedere*, is perhaps the closest approximation in painting to Winckelmann's definition of the central qualities of classical Greek sculpture: 'noble simplicity and calm grandeur' and 'precision of contour'. Although the fresco is painted on a ceiling it is treated as a wall picture and is completely unillusionistic.

putti, medallions, urns, candelabra, tripods and rosettes— were drawn from a mixture of classical sources 'transfused', as Adam himself put it, 'with novelty and variety'. The exquisitely drawn and engraved designs for decorations of this type are among the most satisfying products of this phase of Neo-Classicism. Together with the needle-sharp engravings of Stuart's *Athens*, they mark the opposite extreme of the style from the colossal, Piranesi-inspired architecture of buildings like the Panthéon or old Newgate Prison or the wilder projects of Ledoux.

In France the resistance to the Rococo was more doctrinaire than in other parts of Europe and the neo-classical forms which replaced it were heavier and more severely rectangular. But even here the reaction was not unqualified or truly Greek. Two different areas can be distinguished. In work produced for the court there was some compromise with the Rococo and a return to the style of Louis XIV. Essentially, 'Louis Seize' furniture and decoration depended on the substitution of symmetry, straight lines and simple, mainly abstract ornament for the freely roving curves and naturalistic sprays of the Rococo, as in the tables and cabinets of the *ébéniste*, René Dubois. Similarly, the little round temple in the gardens of the Petit Trianon by the typical 'Louis Seize' architect and successor to Gabriel, Richard Mique, is English in form and Roman in detail and Greek only in the elegance and precision of its treatment.

In Paris, however, there were more serious attempts to be Greek, or at least classical in more than a conventional, decorative sense. As early as 1756–57 the *amateur*, Lalive de Jully, ordered new furniture for his apartment which, according to Cochin, writing in the 1760s, started a fashion for 'garlands looped like well-ropes, clocks in the form of vases, and beautiful inventions which were imitated by all the ignorant and flooded Paris with bits and pieces *à la grecque*.' The furniture that survives shows that it was hardly Greek in the strict sense (as Lalive acknowledged) but it was extremely massive and severe, with hard edges, rectangular forms, ornaments of lions' heads and feet and borders of key and wave pattern design. Going further, one may guess that the 'bits and pieces *à la grecque*' which flooded Paris, influencing not only decoration and furni-

95. **Robert Adam.** *The Sculpture Gallery, Newby Hall, Yorks. c.* 1767–1780. This gallery consists of three rooms built and decorated throughout by Robert Adam to display the sculpture belonging to the owner of the house, William Weddell. The collecting of Antiques was an important feature of the neo-classical period, when English Grand Tourists were the principal buyers. Such collections were not merely curiosities but the inspiration of a way of life. Adam's sculpture gallery at Newby, which is sharp and elegant in detail and beautifully lucid in form, is one of the finest interiors of its date in Europe.

96. **Robert Adam.** *Design for a Ceiling at 7, Queen Street, Edinburgh.* 1770. Pen and watercolour. 18¾ × 18¾ in. (47.5 × 47.5 cm.). Royal Institute of British Architects, London. This design incorporates many of the decorative motives constantly used by Adam; cameos, fans, ribbons, vases, palmettes, acanthus leaves, honeysuckle flowers, and guilloche and indent-pattern borders. It is also a good example of the refined technical precision typical of neo-classical decorative designs.

ture but gold and silver objects, jewellery, textiles and even hairstyles, were culled from such sources as the Herculaneum publications and Caylus's *Recueil d'antiquités*—when they were not simply made up. A characteristic invention of the period was a multi-purpose article of furniture, based on a classical tripod, known as an *Athénienne*. Neo-classical design at this stage was Egyptian, 'Etruscan' and Roman as well as Greek.

We are thus forced to conclude that, despite the enthusiasm for all things Greek, the actual knowledge of Greek art at this period was still limited and confused and the number of works executed in a purely Greek style correspondingly small. In English architecture the main field in which pure Greek design appeared was the garden pavilion; an example is the small Doric temple in the grounds of Hagley Hall, near Birmingham, which was built by Stuart in 1758 and was the first of its kind in Europe. In France the short, stumpy Greek Doric or Tuscan column, rising straight out of the ground without a base, was used in large-scale buildings by architects like de Wailly and Peyre (in the Théâtre de l'Odéon, Paris, 1779–82), Chalgrin, Brongniart and Ledoux. But French architects were influenced by other considerations than Greek revivalism, as must now be explained. The true Greek revival in architecture, which swept across Europe and the United States in the early 19th century, did not begin until the 1790s.

THE CREATION OF 'EXPRESSIVE' ARCHITECTURE

The idea that the design of a building should express its function existed before and has continued to exist since the neo-classical period. But until the 18th century it was not so much the forms themselves as the ways they were used that were expressive. Thus the circular plan and simple mathematical proportions of the ideal Renaissance church symbolised the perfection of God; the dynamic rhythms, contrasting curves and upward soaring movement of the baroque church façade conveyed the power and glory of Catholicism. But in both cases the actual forms used, columns, cornices, pediments and so on—the vocabulary of classical architecture—were irrelevant to the expressive purpose; the same effect might have been gained with forms of quite different, non-classical design. Admittedly there was a belief that each of the orders had its emotional character or 'personality'; it was agreed that the Doric was masculine and the Corinthian feminine, with the Ionic in between. But directly the argument entered into details, confusion and contradiction set in. By the mid-17th century it was questioned whether the so-called characteristics of the orders were really inherent and therefore absolutely true, or whether they were simply established by custom. Certainly by the 18th century the fact that Vitruvius had laid down the general lines of approach was no longer considered a sufficient explanation.

The scepticism towards authority implied in this was the

97. **Josiah Wedgwood.** *Vase with the Apotheosis of Homer. c.* 1789.
Blue and white jasper-ware. 13 in. high (33 cm.). Nottingham
City Art Gallery. This shows the type of elegant vase, adapted
from Greek (then thought to be Etruscan) originals, made by
Wedgwood at his factory called 'Etruria' near Hanley, in
Staffordshire. Jasper-ware was invented about 1774–5.
The graceful lines and the forms of the decoration are akin to
Robert Adam's interiors. The very pure outlines of the figurative
relief, designed by the sculptor, John Flaxman, and the
Homeric subject-matter, are also typical of Neo-Classicism.

98. **Louis Joseph le Lorrain.** *Filing Cabinet. c.* 1756–57.
Oak veneered with ebony, with decorative motives in gilt
bronze. 63½ × 42½ in. (161 × 108 cm.). Musée Condé,
Chantilly. This cabinet formed part of a suite of furniture made
for the Parisian *amateur*, Ange-Laurent de Lalive de Jully,
who was clearly the inspirer of the design. Completely different
in style from rococo furniture, it was called *à la grecque*, although
the owner himself realised that it was not Greek in the true sense.
Its main features are its heavy, square-edged forms, straight
lines and abstract classical ornament. The looped garlands at
the top were nick-named well-ropes *(cordes de puits)*.

starting point of the new, revolutionary theories of Cordemoy (1706) and the Venetian teacher, Lodoli, which together were summed up and developed further by Laugier
in his *Essai sur l'architecture* of 1753. Not that these architectural philosophers wanted to discard the expressionist
approach to architecture and see its forms governed purely
by aesthetics. On the contrary, what they had in mind was
the rejection of the aesthetic approach; the orders were no
longer to be used ornamentally or as wall decoration, but
solely in accordance with their function. The most radical
definition of this function was given by Laugier.

Laugier based his argument on Vitruvius's contention
—accepted ever since as a historical curiosity—that the
Doric order was originally developed from the uprights and
cross-beams of wooden temples. He then went further and
deduced that the wooden temple must have been derived
from a still earlier construction, the primitive hut. Like the
appeal to 'Etruscan' sources in the debate between the
supporters of Greek and Roman architecture, this was another example of that fascination with origins that marked

the 18th century. It was equally an example of the fascination with nature as the ultimate authority. The primitive
hut preceded all other forms of construction and can only
have been evolved from a natural source. It was also
necessarily functional and therefore rational.

Modern architecture, argued Laugier, should similarly
return to first principles. It too should be rational and
functional. And just as the primitive hut, as Laugier
visualised it (he was the first to do so) would have consisted
only of upright posts, cross-beams and a pitched roof, so the
ideal modern building should consist entirely of the same
basic forms—columns, cross-beams and a roof. This was
the final challenge to Vitruvius's authority and to the whole
later development of classical architecture. What Cordemoy had already called 'architecture in relief'—pilasters,
half-columns, ornamental pediments, attic storeys and so
on—all those forms previously used to articulate a wall
surface—should go. Laugier even disagreed with Vitruvius
about the proportions and number of the orders; he saw no
reason why these should be sacrosanct.

99 (left). **Claude-Nicolas Ledoux.**
Barrière de la Villette (or *de St Martin*),
Paris. 1784–89. This is one of four toll-
houses surviving out of twenty or so by
Ledoux built on the outskirts of Paris just
before the Revolution. The massive forms
reflect the function of the building—to
enforce the payment of taxes on grain
brought into Paris—and exemplify
Ledoux's belief that architectural beauty
should depend only on the use of pure
geometrical form.

100 (opposite). **Jacques-Louis David.**
*The Lictors bringing back to Brutus the Bodies
of his Sons.* 1789. Oil on canvas. 128 × 164
in. (325 × 423 cm.). Louvre, Paris. Shown
in Paris after the outbreak of the
Revolution, this work directly reflects
events. Brutus, suppressing private feelings
in the interests of the State, turns his back
on the bodies of his sons, executed for
treason on his own order. Earlier neo-
classical paintings had been crude; this is a
masterpiece—heroic, authoritative,
impeccably composed.

Laugier's influence was profound and was greater than appears at first sight, for it is obvious that few modern versions of the primitive hut were required. His contribution was to assert that the forms of a building should proclaim its function and, although he passionately believed that the only proper form of upright was the column not the wall, he opened the way to the use of forms of any shape, arrangement or size. Visual beauty was to arise only from simple geometrical shapes and to be subordinate to character and expression. From this followed the tough, 'brutalist' quality of some neo-classical architecture, especially in France.

One of the first buildings to reflect Laugier's influence was the Panthéon by Soufflot, although, as we have seen, it also owed something to Piranesi and Roman architecture and depended on a 17th-century source—Wren's St Paul's Cathedral—for its dome. Originally the design would have been closer to Laugier's ideal for Soufflot wanted to make the outside 'more window than wall' but for structural reasons the windows had to be filled in. Inside he did succeed in supporting the roof, including the dome, entirely on columns, not piers (it is worth noting that he was also an admirer of Gothic construction). The detail throughout is of extraordinary, chilling beauty—very restrained, diamond-hard and austerely inventive, especially at the back. Needless to say it includes no ornamental or figurative sculpture. Appropriately, the building was praised by Laugier as 'the first example of a perfect architecture'.

Probably the most important building to show Laugier's influence in England before the time of Sir John Soane was Dance's Newgate Prison, (designed in 1769 and demolished in 1906); Laugier's functionalism was there combined with the melodrama of Piranesi to produce a building of consciously frightening power. In France the chief exponent of this approach was Claude-Nicolas Ledoux. Ledoux was the type of the 18th-century philosophical architect, as Laugier was the century's typical architectural philosopher. He summed up the rationalist tendencies

of the age and, carrying them to their logical conclusion, undermined all its traditional methods. His favourite forms were the pyramid, the cylinder and the cube, and he sometimes dispensed entirely with classical motives.

Many of his works were purely theoretical and only existed in engraved form in a treatise published towards the end of his life, in 1804. They included a project for an ideal city, which looked forward to the secular, functional Utopias of the 20th century rather than backward to the symbolic Cities of God of the Renaissance. It was also typical of Ledoux's modern brand of functionalism to devise appropriate houses not merely for different classes of people but for people of different professions, activities and states of life; they included the House for the Writer, the House for the Broker, the Shelter for the Rural Guards, the House for Four Families, etc.. Moreover, although his buildings were formally composed of related units of solid geometry, without ornament of any kind, they were not abstract. On the contrary, their design was consciously expressive of their purpose, often in a quasi-symbolic rather than a practical sense.

The toll-houses or *barrières* erected between 1784 and 1789 on the outskirts of Paris exemplify this. Their purpose was the collection of a newly imposed tax on grain brought into the city from the countryside. Looking at, for example, the Barrière de St Martin, it is hard to avoid the feeling that Ledoux enjoyed his task. Although he was a revolutionary architect and dabbled in ideas about public morals and law which have a Rousseauesque flavour, he was an authoritarian in politics and, until the Revolution, a loyal servant of the government (he later became an admirer of Napoleon). His toll-houses were surely larger than necessary for practical purposes, and their massive, square-edged and brutal forms must have had an intentionally demoralising effect on the unfortunate peasants who had to pay taxes at them. It is not surprising that some were torn down by the mob during the Revolution or that their architect narrowly escaped the guillotine.

Despite Ledoux's conservative position in politics compared with that of Jacques-Louis David, there were interesting parallels between them in temperament and artistic personality. Both were to some extent *philosophes*, yet both were animated as much by passion as by reason. In treating the *Oath of the Horatii* as a pre-Revolutionary manifesto, David chose a subject whose implications were as eloquent of the subjection of the individual to the good of the state as were Ledoux's toll-houses. In *The Lictors bringing back to Brutus the Bodies of his Sons*, exhibited a few weeks after the start of the Revolution, this theme became explicit. Brutus who had condemned his sons to death for treachery, stoically turned his back on their bodies when they were brought back from execution.

But apart from these considerations of subject-matter, there is a Ledoux-like toughness and austerity in David's early style which sharply differentiates his interpretation of Neo-Classicism from the effete reworking of the Antique recommended by Winckelmann. In feeling and style his art looked back in many ways to Poussin; he used the same relationship of figures to space, the same method of modelling forms by light and shade and the same type of compositional arrangement. No other artist but Poussin created the same tension between the severity of the forms and the depth of emotion they contain. Yet there was also a new quality, for David's paintings expressed a modern, personal kind of heroism that was to be identified forty years later as Romantic by Stendahl.

CONCLUSION

What are the main themes that have been surveyed in this book? On a general view, perhaps the most striking is the triumph of aristocratic art in the broadest sense, meaning by that, art of all kinds designed for the enjoyment of royal and noble patrons. Art of this sort reached its zenith in the age of Baroque and dominated that age as it dominated no other. A second typical development was the rise of por-

traiture, landscape, *genre* painting and interior decoration —the previously 'lesser' categories—to new levels of importance. Other features might be described as 'lasts' rather than 'firsts'. For the last time in history the Church was a leading patron of the best available art; in fact it was already losing this role before the period ended. Also for the last time, the artist was still an integrated member of society, not at odds with it (apart from rare exceptions) as he was to be in the 19th century.

More than this, art itself revolved for the last time round its relationship to the Antique. It was also a more complicated relationship than ever before, ranging from complete subservience to classical ideals at one extreme to complete indifference to them at the other, with almost all possible shades existing in between. The end of the period was the most complex of all from this point of view. In some quarters old ideals were being re-affirmed more rigorously than ever; in others, new attitudes, concerning not only styles but the nature of art itself, were finding expression.

In their various ways both Hogarth's *Analysis of Beauty* (1753) and Diderot's *Salon* criticisms were attempts to discover new ways of discussing art. Other examples are Burke's treatise on the *Sublime and the Beautiful* (1757), Gilpin's essays on *Picturesque Beauty* (1770s onwards) and Kant's writings on aesthetics, all of which laid a new emphasis on the subjective experience of the observer in discussing the nature of art. None, incidentally, was written from within the fold of academies and none continued to uphold the old hierarchy of categories.

Despite its preoccupation with Antiquity, the neo-classical period was the first since the Middle Ages to formulate aesthetic theories that freed art from the classical tradition. This time the release was permanent; later classical revivals, as in the early 20th century, have been isolated eddies which formed themselves temporarily in an otherwise anti-classical current.

Biographical Notes on Artists and Writers on Art of the Baroque Period mentioned in this book

Adam, Robert. Scottish architect, interior designer; *b.* 1728 Kirkcaldy (Fife); visited France, Italy, the Rhine, 1754–8; worked throughout England and Scotland; *d.* London 1792.

Agucchi, Mons Giovanni Battista. Italian theorist and patron; *b.* 1570 Bologna, lived mainly Rome, where *d.* 1632; friend of Domenichino, Guercino, secretary to Pope Gregory XV (1621–3).

Algardi, Alessandro. Italian sculptor; *b.* 1595 Bologna; worked in Rome from *c.* 1625, where *d.* 1654.

Asam, Cosmas Damian. German architect, painter, decorator; *b.* 1686 Benediktbeuren (Bavaria); studied Rome 1711–13; worked chiefly Bavaria and Tyrol; *d.* Weltenburg 1739.

Asam, Egid Quirin. German architect, sculptor, brother of preceding, with whom collaborated; *b.* 1692 Benediktbeuren (Bavaria); studied Rome 1711–13; worked chiefly Bavaria and Tyrol; *d.* Munich 1750.

Audran, Claude (called Claude III Audran). French decorative artist, designer of arabesques and tapestries; *b.* 1658 Lyon, worked Paris, where *d.* 1734.

Bähr, Georg. German architect; *b.* 1666 near Lauenstein (Saxony); worked in and near Dresden, where *d.* 1738.

Barry, James. English figure painter, etcher; *b.* 1741 Cork (Ireland); arrived London 1764, studied Rome 1766–71, then worked London, where *d.* 1806.

Batoni, Pompeo. Italian portrait and figure painter; *b.* 1708 Lucca; worked Rome probably from 1728, where *d.* 1787.

Bellori, Giovanni Pietro. Italian archaeologist and writer on art; *b. c.* 1615 Rome, where lived and *d.* 1696; friend of Poussin.

Bernini, Gianlorenzo. Italian sculptor, architect, painter, stage designer; *b.* 1598 Naples; arrived Rome 1605, where worked and *d.* 1680; visited Paris 1665.

Bibiena, Giuseppe Galli. Italian stage designer, theatre architect; son and brother of artists in same profession; *b.* 1696 Parma; worked Vienna, Dresden, Prague, Warsaw, N. Italy, etc.; *d.* Berlin 1757.

Boffrand, Germain. French architect, interior designer; *b.* 1667 Nantes; arrived Paris *c.* 1681, where mainly worked and *d.* 1754; visited Würzburg 1724.

Borromini, Francesco. Italian architect; *b.* 1599 Bissone on Lake Lugano, kinsman of Maderna; arrived Rome *c.* 1614 from Milan; worked Rome, where *d.* 1667.

Bossi, Antonio. Italian stucco-worker, member of family in same profession; *b.* before 1620 Porto on Lake Lugano; possibly trained Vienna; worked Würzburg 1735–53, where *d.* 1764.

Bouchardon, Edmé. French sculptor; *b.* 1698 Chaumont-en-Bassigny (Haute-Marne); arrived Paris 1721; in Rome 1723–32, then worked mainly Paris, where *d.* 1762.

Boucher, François. French painter, draughtsman, decorator, engraver; *b.* 1703 Paris; in Italy 1727–31, then worked Paris, where *d.* 1770.

Braun, Mathias. Bohemian sculptor; *b.* 1684 Mühlau (Tyrol); probably trained Italy; arrived Prague 1709/10, where mainly worked and *d.* 1738.

Brongniart, Alexandre-Théodore. French architect; *b.* 1739 Paris, where worked and *d.* 1813.

Brosse, Salomon de. French architect; *b.* 1571 Verneuil-sur-Oise; from 1598 worked in and near Paris, where *d.* 1626.

Brouwer, Adriaen. Flemish genre painter; *b.* 1605/6 Oudenaarde; in Amsterdam by 1625, then Haarlem; by 1631/2 back in Antwerp, where *d.* 1638.

Burlington, Richard Boyle, 3rd Earl of. English patron and architect; *b.* 1694; visited Italy 1714–15 and 1719; lived mainly London, *d.* 1753.

Bustelli, Franz Anton. Swiss sculptor in porcelain; *b.* 1723 Locarno; from 1754 chief designer and modeller at Nymphenburg porcelain manufactory, near Munich, where *d.* 1763.

Caffieri, Jacques. French sculptor, bronze-worker; *b.* 1678 Paris, of Italian father; worked Paris, where *d.* 1755. In later years assisted by son, Philippe (1714–74).

Campbell, Colin (or Colen). Scottish architect; active London and other parts of England from 1712; *d.* London 1729.

Canaletto (Antonio Canale) Italian view painter, draughtsman, etcher; *b.* 1697 Venice; visited Rome 1719–20, England 1746–*c.* 1756 (with interruptions); otherwise worked Venice, where *d.* 1768.

Caravaggio, Michelangelo Merisi da. Italian figure painter; *b.* 1573 Caravaggio (Lombardy); arrived Rome *c.* 1592 from Milan; in Rome until 1606, then Naples, Malta, Sicily; *d.* Port' Ercole (Tuscany) 1610.

Carracci, Annibale. Italian figure, portrait, landscape, genre painter and draughtsman; *b.* 1560 Bologna, where worked until 1595; visited Parma, Venice 1586–7; in Rome from 1595 where *d.* 1609.

Carracci, Ludovico. Italian figure painter and draughtsman; cousin, teacher and collaborator of preceding; *b.* 1555 Bologna, where worked and *d.* 1619; visited Rome 1602.

Casas y Novoa, Fernando de. Spanish architect; active 1711 – *d.* 1794; worked Santiago.

Cavallino, Bernardo. Italian figure painter; *b.* 1616 Naples, where worked and *d.* 1656.

Caylus, Anne-Claude-Philippe de Tubières, Comte de. French archeologist, engraver, writer on art; *b.* 1692 Paris; began career as a soldier, then travelled Italy, Greece; afterwards lived Paris, where *d.* 1765.

Chalgrin, Jean-François Thérèse. French architect; *b.* 1739 Paris, where mainly worked and *d.* 1811.

Chambers, Sir William. English architect; *b.* 1723 Göteborg (Sweden) of Scottish parents; visited China before studying architecture Paris, 1749, and Italy, 1750–55; then worked London, where *d.* 1796.

Chambray, Roland Fréart de. French connoisseur, writer on art and architecture; *b.* 1606 Paris, where lived and *d.* 1676; friend of Poussin.

Champaigne, Philippe de. French figure and portrait painter; *b.* 1602 Brussels; arrived Paris 1621, where worked and *d.* 1674.

Chantelou, Paul Fréart de. French connoisseur, patron and diarist, brother of Fréart de Chambray; *b.* 1609 Paris, where lived and *d.* 1694; visited Rome 1640–3; friend and patron of Poussin, guide to Bernini in Paris, 1665.

Chardin, Jean-Baptiste-Simeon. French genre and still-life painter; *b.* 1699 Paris, where worked and *d.* 1779.

Claude Lorrain (properly Claude Gellée). French landscape painter and draughtsman; *b.* 1600 Chamagne (Lorraine); in Rome by 1620, where worked and *d.* 1682.

Clodion (properly Claude Michel). French sculptor; *b.* 1738 Nancy; arrived Paris 1755, where worked and *d.* 1814; visited Rome 1762–71.

Cochin, Charles-Nicolas, the Younger. French painter, engraver and writer on art; *b.* 1715 Paris, where worked and *d.* 1790; visited Italy 1749–51 with Soufflot and Marigny; historical secretary to French Academy of Painting and Sculpture from 1755.

Cordemoy, Abbé Louis-Gérard de. French theologian, historian and writer on architecture; *b.* 1651 Paris, where lived and *d.* 1722.

Cortona, Pietro Berrettini da. Italian architect, figure painter and decorator; *b.* 1596 Cortona; in Rome by 1620, where mainly worked and *d.* 1669; visited Florence 1637 and 1640–6.

Coysevox, Antoine. French sculptor; *b.* 1640 Lyon; arrived Paris 1657, where worked and *d.* 1720; also active at Versailles.

Cressent, Charles. French cabinet-maker and sculptor, pupil of Coysevox; *b.* 1685 Amiens, worked mainly Paris, where *d.* 1768.

Cuvilliés, François. Franco-German architect and interior designer; *b.* 1695 Soignies (Hainaut, Flanders); in Munich at Bavarian Court from *c.* 1708; studied Paris 1720–4, then worked chiefly in and near Munich, where *d.* 1768.

Cuyp, Albrecht (or Aelbert). Dutch landscape and portrait painter; *b.* 1620 Dordrecht, where chiefly worked and *d.* 1691.

Dance, George, the Younger. English architect, son of George Dance the Elder, also architect; *b.* London 1741, where worked and *d.* 1825; visited Italy 1758–65.

David, Jacques-Louis. French figure and portrait painter; *b.* 1748 Paris; visited Rome 1775–81, then worked Paris until 1815, afterwards living in Switzerland and Brussels, where *d.* 1825.

Diderot, Denis. French man of letters, encyclopedist and art critic; *b.* 1713 Langres; lived mainly Paris; visited Russia 1773; *Salon* criticisms periodically from 1759–79; *d.* Paris 1784.

Dientzenhofer, Johann. German architect; *b.* 1663 near Aibling; trained Prague; visited Italy 1699–1700; worked Central Germany; *d.* Bamberg 1726.

Domenichino (properly Domenico Zampieri). Italian figure and landscape painter and draughtsman; *b.* 1581 Bologna; arrived Rome 1602, where mainly worked until 1630, then moved to Naples, where *d.* 1641.

Dubois, René. French cabinet-maker, son and brother in same profession; *b.* 1737 Paris, where worked and *d.* 1799; patronised by Marie Antoinette.

Dughet, Gaspard. French landscape painter and draughtsman; *b.* 1615 Rome of French father; pupil (1631–5) of Nicolas Poussin, whose name he adopted (hence sometimes called Gaspard Poussin); worked Rome, where *d.* 1675.

Duplessis, Jean-Claude (properly Jean-Claude Chamberlan). French sculptor, son of Claude-Thomas Chamberlan (also called Duplessis), who was goldsmith and designer of bronze mounts for Sèvres porcelain from 1747; J.-C. Duplessis active as modeller at Sèvres from 1761 until *d.* 1783.

Duquesnoy, François. Flemish sculptor; *b.* 1594 Brussels; worked in Rome from 1618; *d.* 1643 on way to Paris.

Effner, Joseph. German architect and landscape gardener; *b.* 1687 Dachau (Bavaria); visited Paris 1706–15, where studied under Boffrand; afterwards worked Munich, *d.* 1745.

Elsheimer, Adam. German figure and landscape painter and draughtsman; *b.* 1578 Frankfurt; in Venice 1598/9; arrived Rome 1600, where worked and *d.* 1610.

Falconet, Etienne-Maurice. French sculptor; *b.* 1716 Paris; pupil of J. B. Lemoyne; worked Paris, where *d.* 1791; Director of Sculpture at Sèvres 1757–66; visited Russia 1766–79.

Feichtmayr, Johann Michael (or Feuchtmayr, Feichtmair, etc.). German sculptor and stucco-worker, member of large family in same profession; *b.* 1709/10 near Wessobrunn, Upper Bavaria; worked Bavaria and Central Germany; *d.* Augsburg 1722.

Fernandez, Gregorio. Spanish sculptor; *b.* 1576 Galicia; worked Valladolid, where *d.* 1636.

Fischer, Johann Michael. German architect; *b.* 1692 Oberpfalz district of Bavaria; visited Moravia 1715–16, afterwards worked Bavaria; *d.* Munich 1766.

Fischer von Erlach, Johann Bernhard. Austrian architect; *b.* 1656 Graz; visited Rome and Naples *c.* 1674–87; in Vienna by 1690, where mainly worked *d.* 1723.

Flaxman, John. English sculptor; *b.* 1755 York, but family soon moved London, where Flaxman mainly worked and *d.* 1826; associated with Wedgwood from 1775; visited Italy 1787–94.

Fontana, Carlo. Italian architect; *b.* 1634 near Como; arrived Rome by 1655, where worked and *d.* 1714.

Fragonard, Jean-Honoré. French figure and landscape painter; *b.* 1732 Grasse, pupil of Chardin and Boucher in Paris; visited Italy 1756–61 and 1773; otherwise worked mainly in Paris until Revolution and *d.* there 1806.

Gabriel, Ange-Jacques. French architect; *b.* 1698 Paris; worked Fontainebleau, Versailles, Paris, where *d.* 1782.

Gainsborough, Thomas. English portrait and landscape painter; *b.* 1727 Sudbury (Suffolk); in London from 1740; *c.* 1748–59 mainly at Ipswich; 1759–74 at Bath, then London, where *d.* 1788.

Gaudreaux, Antoine-Robert. French cabinet-maker to Louis XV; *b. c.* 1680, worked Paris, where *d.* 1751.

Gaulli, Giovanni Battista, called Bacciccia. Italian figure and portrait painter; *b.* 1639 Genoa; arrived Rome very young and worked and *d.* 1709.

Gibbs, James. Scottish architect; *b.* 1682 near Aberdeen; travelled through Central Europe to

Rome, where studied under Fontana and stayed 1703–9; afterwards worked mainly Oxford, Cambridge and London, where d. 1754.

Gillot, Claude. French genre painter, draughtsman, engraver and stage designer; b. 1673 Langres (Champagne); worked Paris, where d. 1722.

Gilpin, Rev. William. English draughtsman, tourist and writer on the Picturesque; b. 1724 near Carlisle; Oxford 1740–6; lived as country vicar in Hampshire from 1777 until d. 1804.

Girardon, François. French sculptor; b. 1628 Troyes; visited Rome 1645–50, then worked mainly Versailles and Paris, where d. 1715.

Goya (properly Francisco de Goya y Lucientes). Spanish figure and portrait painter, draughtsman, etcher and tapestry designer; b. 1746 Saragossa; arrived Madrid 1766; visited Rome 1771; otherwise worked mainly Madrid until 1824, when visited Paris; d. Bordeaux 1828.

Greco, El (properly Domenikos Theotocopoulos). Spanish figure and portrait painter; b. 1541 Crete; in Venice probably from 1558–76, interrupted by visit to Rome 1570–2; in Toledo, Spain, by 1577, where worked and d. 1614.

Greuze, Jean-Baptiste. French genre painter; b. 1725 Tournos (Saône-et-Loire); visited Italy 1755–6; otherwise worked Paris, where d. 1805.

Guardi, Francesco. Italian view painter and draughtsman; b. 1712 Venice, where worked and d. 1793.

Guarini, Guarino. Italian architect and mathematician, member of Theatine Order; b. 1624 Modena; arrived Rome 1639; in Modena 1647–55, Messina 1660–6; worked Turin 1666 until d. 1683; also visited Paris (1662), Prague and possibly Spain and Portugal.

Guercino (properly Francesco Barbieri). Italian figure painter; b. 1591 Cento (Emilia); visited Bologna before 1620, Rome 1621–3; afterwards worked Cento and Bologna, where d. 1666.

Günther, Ignaz. German sculptor; b. 1725 near Ingolstadt (Bavaria); studied Munich 1743–50, also Salzburg, Mannheim and Vienna (1750–52); afterwards worked Munich, where d. 1775.

Hals, Frans. Dutch portrait painter; b. 1580/5 probably Antwerp; in Haarlem by 1591, worked and d. there 1666.

Hamilton, Gavin. Scottish figure painter, archaeologist and dealer; b. 1723 in Lanarkshire; from 1748 lived mainly in Rome, though sent paintings to London; d. there 1798.

Hawksmoor, Nicholas. English architect; b. 1661 in Nottinghamshire; arrived in London by 1680, when became assistant to Wren; from 1705 associated with Vanbrugh, but also important architect in his own right; d. London 1736.

Heda, Willem Claesz. Dutch still-life painter; b. 1594 Haarlem, where worked and d. 1682.

Hildebrandt, Lukas von. Austrian architect and military engineer; b. 1663 Genoa of German parents; visited Rome c. 1690, where studied under Fontana; in Piedmont 1695–6; afterwards worked mainly in and near Vienna, where d. 1745; also sent designs to Central Germany (Würzburg, Pommersfelden) and Bohemia.

Hogarth, William. English genre, portrait and figure painter and engraver; b. 1697 London, where worked and d. 1764; visited France 1743 and 1748.

Hooch (or Hoogh), **Pieter de.** Dutch genre painter; b. 1629 Rotterdam; in Delft by 1654; moved to Amsterdam by 1663, where worked and d. after 1684.

Houdon, Jean-Antoine. French sculptor; b. 1741 Versailles; visited Rome 1764–8; otherwise worked Paris, where d. 1828.

Jones, Inigo. English architect and stage and costume designer; b. 1573 London, where mainly worked and d. 1652; visited Italy 1601 (?), c. 1605 and 1612–15; also visited France.

Jorhan, Christian, the Elder. German sculptor; b. 1727 Griesbach (Bavaria); trained Munich and influenced by Günther; settled in Landshut (Bavaria) 1775 and d. there 1804.

Juvarra (or Juvara), **Filippo.** Italian architect; b. 1678 Messina; in Rome 1703–14, where studied under Fontana; worked in and near Turin from 1714; arrived 1735 Madrid, where d. 1736.

Kalf, Willem. Dutch still-life painter; b. 1619 Amsterdam; worked Rotterdam, Paris and Amsterdam, where d. 1693.

Kändler, Johann Joachim. German sculptor in porcelain; b. 1706 near Bischofswerda (Saxony); in Dresden from 1723; head modeller and designer at Meissen porcelain manufactory from 1731; d. Meissen 1775.

Kauffmann, Angelica. Anglo-Swiss figure and portrait painter; b. 1741 Coire (Switzerland); visited Italy as child and until arrived London, 1766; returned Italy 1781; d. Rome 1807.

Kent, William. English architect, painter, interior designer and landscape gardener; b. 1685(?) in Yorkshire; studied painting in Rome 1709–19, then lived mainly London, where d. 1748, having worked throughout England; protégé and friend of Burlington.

Kneller, Sir Godfrey. Anglo-German portrait painter; b. 1646/9 Lübeck (Germany); studied Amsterdam and Italy before arriving London, 1674, where worked and d. 1723.

Knobelsdorff, Georg Wenceslaus von. German architect, interior designer and painter; b. 1699 Kr. Crossen (Oder). Began career in army, then trained as painter (1729), visited Italy 1736–7, Dresden and Paris, 1740; worked for Frederick the Great at Potsdam and Berlin, where d. 1759.

Koninck (or Koninck), **Philips.** Dutch landscape painter; b. 1619 Amsterdam, where mainly worked and d. 1688.

Krämer (or Kramer), **Simpert.** German master mason from Edelstetten (Swabia); dates of birth and death unknown; active 1717–57, chiefly at Ottobeuren.

Lalive de Jully, Ange-Laurent de. French diplomat, amateur and collector; Master of Ceremonies at Versailles; b. 1725 Paris, where mainly lived and d. 1775.

Largillière, Nicolas de. French portrait painter; b. 1656 Paris, but brought up in Antwerp; c. 1674–82 in London as assistant to Lely; afterwards worked Paris, where d. 1746.

La Tour, Georges de. French figure painter; b. 1593 Vic-sur-Seille (Lorraine); possibly visited Rome and/or Holland c. 1620; otherwise worked Lunéville (Lorraine), where d. 1652.

La Tour, Maurice Quentin de. French portraitist in pastel; b. 1704 St Quentin (Aisne); visited Paris and London when young; worked Paris 1724–84; d. St Quentin 1788.

Laugier, Abbé Marc-Antoine. French preacher, man of letters and architectural theorist; b. 1713 Manosque (Provence); Jesuit priest until 1754, when left the Order; lived mainly Paris from 1744 and d. there 1769.

Lebrun (or Le Brun), **Charles.** French figure painter and decorator; b. 1619 Paris, where pupil of Vouet 1634–7; in Italy 1642–6, where met Poussin; afterwards worked Versailles and Paris, where d. 1690; Director of French Academy of Painting and Sculpture from 1663, though effective power ceased after death of Colbert in 1683.

Ledoux, Claude-Nicolas. French architect and writer on architecture; b. 1736 Dormans (Marne); worked chiefly in and near Paris, where d. 1806.

Lely, Sir Peter. Anglo-Dutch painter; b. 1618 Soest (Westphalia) of Dutch parents; trained Haarlem; probably arrived London by 1643, where worked and d. 1680.

Lemoyne, Jean-Baptiste, the Younger. French sculptor, son of J.-B. Lemoyne the Elder, also a sculptor; b. 1704 Paris, where worked and d. 1778.

Le Nain, Louis. French genre painter, brother of Antoine (c. 1588–1648) and Mathieu Le Nain (c. 1607–77); b. 1593 Laon; possibly studied Rome; in Paris by 1630, where worked and d. 1648.

Lepautre, Pierre. French interior designer, draughtsman and engraver, son of engraver Jean Lepautre (1618–82) and cousin of sculptor of same name as his own (1660–1744); b. c. 1648 Paris, where worked—also at Versailles—and d. c. 1716.

Le Roy, Philbert. French architect; active Paris 1623–44.

Le Vau, Louis, the Elder. French architect; b. 1612 Paris, where worked—also at Versailles—and d. 1670.

Lodoli, Carlo. Italian teacher of architectural theory; priest in Franciscan Order; b. 1690 Venice where lived and died 1761.

Longhena, Baldassare. Italian architect; b. 1598 Venice; worked in and near Venice and d. there 1682.

Lorrain, Louis-Joseph le. French view-painter and cabinet-maker; b. 1715 Paris, where mainly worked; 1758 visited St Petersburg where d. 1759.

Maderna, Carlo. Italian architect; b. 1556 near Lake Lugano; arrived Rome by 1585, where worked and d. 1629; chief architect to St Peter's from 1603.

Mansart, François. French architect; b. 1598 Paris; worked in and near Paris and d. there 1666.

Mansart, Jules-Hardouin. French architect, great nephew of preceding; b. 1646 Paris; worked chiefly in and near Paris and, from 1678, at Versailles; Surveyor of Royal Works from 1699; d. Paris 1708.

Marchione (or Marchionni), **Carlo.** Italian architect; b. 1702 Rome, worked and d. there 1786.

Meissonnier, Juste-Aurèle. French designer of gold and silver ornaments; b. 1695 Turin of Provençal parents; in Paris by 1723, where worked and d. 1750.

Mengs, Anton Raphael. German figure and portrait painter; b. 1728 Aussig (Bohemia); first visited Rome 1741, where mainly worked; also worked Dresden 1744–59 and Spain from 1761; d. Rome 1779.

Mique, Richard. French architect and designer of garden ornaments; b. 1728 Nancy (Lorraine), where mainly worked until 1766; afterwards active chiefly at Versailles; d. (guillotined) Paris 1794.

Montáñez, Juan Martinez. Spanish sculptor; b. 1568 Alcala la Real; trained Granada 1579–82(?); settled Seville by 1588 and worked and d. there 1649.

Murillo, Bartolomé Esteban. Spanish figure painter; b. 1617 Seville, where mainly worked and d. 1682; visited Madrid c. 1648–51.

Natoire, Charles-Joseph. French figure painter and tapestry designer; b. 1700 Nimes; studied Rome 1723–9; worked mainly Paris 1737–51, when made Director of French academy at Rome; d. Castelgandolfo near Rome 1777.

Nattier, Jean-Marc. French portrait painter; b. 1685 Paris; visited Amsterdam 1717; worked and d. Paris 1766.

Neumann, Balthasar. German architect and military engineer; b. 1687 Eger (Hungary); studied Würzburg from 1711; visited Milan and Vienna, 1718, and Paris, 1723; worked mainly Central Germany (Würzburg, Vierzehnheiligen, Bruchsal, Trier, etc.); d. Würzburg 1753.

Nollekens, Joseph. English sculptor; b. 1737 London; in Rome 1759–70; afterwards worked London, where d. 1823.

Oeben, Jean-François. French cabinet-maker; b. c. 1720 Ebern (Franconia) of German parents; arrived Paris young and worked and d. there 1763.

Oppenord, Gilles-Marie. French interior designer; b. 1672 Paris of Dutch father; studied Rome 1692–9; afterwards worked Paris, where d. 1742.

Pannini, Giovanni Paolo. Italian view painter; b. c. 1692 Piacenza; arrived Rome 1715, where worked and d. 1765.

Patel, Pierre, the Elder. French landscape painter; b. c. 1605, probably Picardy; in Paris by 1635, where worked and d. 1676; pupil of Vouet and influenced by Claude Lorrain.

Peña de Toro, José. Spanish architect; active Santiago from 1652, having arrived from Salamanca; d. 1676.

Perrault, Claude. French architect, doctor and engineer; b. 1613 Paris, where worked and d. 1688.

Perronneau, Jean-Baptiste. French portraitist in oils and pastel; b. 1715 Paris; from c. 1755 increasingly forced by competition to work in provinces and abroad; visited Rome (1759), London (1761) and St Petersburg (1781); d. Amsterdam 1783.

Peyre, Joseph. French painter and architect; b. 1730 Paris; studied Rome 1753–7; afterwards Paris where d. 1788; collaborated with Wailly.

Pineau, Nicolas, French decorative sculptor; b. 1684 Paris; in Russia 1716–27(?); by 1732 back in Paris, where d. 1754.

Piranesi, Giovanni Battista. Italian etcher of ruins and architectural views, architect and draughtsman; *b.* 1720 Venice; arrived Rome 1740, where worked and *d.* 1778.

Pöppelmann, Matheas Daniel. German architect; *b.* 1662 Herford (Westphalia); arrived Dresden, *c.* 1686, where chiefly worked and *d.* 1736; visited Vienna and Rome 1710.

Porcellis, Julius. Dutch marine painter; *b. c.* 1609 Rotterdam (?); *d.* Leyden 1645. Son and pupil of Jan Porcellis (1584–1632).

Poussin, Nicolas. French figure and landscape painter and draughtsman; *b.* 1594 in Normandy; arrived Paris *c.* 1612; lived in Rome *c.* 1624 onwards, except visit to Paris, 1640–42; *d.* Rome 1665.

Pozzo, Padre Andrea. Italian figure painter and decorator; *b.* 1642 in Rome 1681–1702, afterwards visited Vienna and *d.* there 1709.

Prandtauer, Jakob. Austrian architect; *b.* 1660 Stanz (Tyrol); worked throughout Austria; *d.* St Pölten, Lower Austria, 1726.

Preti, Mattia. Italian figure painter; *b.* 1613 Calabria; arrived Rome *c.* 1630; also worked Naples and Malta, where *d.* 1699.

Puget, Pierre. French sculptor and decorator of ships; *b.* 1620 Marseilles; 1640–3 studied Rome, Florence; otherwise worked Genoa, Toulon and Versailles; *d.* 1694.

Rainaldi, Carlo. Italian architect; *b.* 1611 Rome, where worked and *d.* 1691.

Ramsay, Allan. Scottish portrait painter; *b.* 1713 Edinburgh; visited Rome 1736–8, when pupil of Solimena, and again 1754–7; 1775–7 and 1782–4, also visited France 1765; otherwise worked mainly London; *d.* 1784 at Dover on way back from Rome.

Rembrandt Harmensz van Rijn. Dutch figure, portrait and landscape painter, etcher and draughtsman; *b.* 1606 Leyden; trained Amsterdam 1624; settled Amsterdam 1631/2, where worked and *d.* 1669.

Reni, Guido. Italian figure painter; *b.* 1575 near Bologna; visited Rome at various times 1601–21 and Naples 1622; otherwise worked Bologna, where *d.* 1642.

Revett, Nicholas. English architect and writer on architecture; *b.* 1720 Suffolk; visited Naples 1748 and Athens 1751; otherwise lived England; *d.* London 1804.

Reynolds, Sir Joshua. English portrait painter and writer on art; *b.* 1723 Plympton (Devon); in London from 1740, where *d.* 1792; visited Italy 1749–52; President of Royal Academy from 1768.

Ribera, Jusepe de. Spanish figure painter and etcher; *b.* 1591 near Valencia; settled Naples by 1616, where worked and *d.* 1652.

Ricchino, Francesco Maria. Italian architect; *b.* 1583 Milan; studied Rome *c.* 1602–7; afterwards in and near Milan, where *d.* 1658.

Ricci, Marco. Italian landscape and 'caprice' painter, nephew of the following, with whom collaborated; *b.* 1676 Venice; visited Florence, possibly Milan and Rome, and certainly England (1708–10, 1712–16); otherwise worked Venice, where *d.* 1730.

Ricci, Sebastiano. Italian figure painter, uncle of the preceding; *b.* 1660 Belluno (near Venice); worked throughout Italy and Vienna before visiting England with nephew, 1712–16; returned via Paris to Venice, where *d.* 1734.

Robert, Hubert. French landscape and 'caprice' painter; *b.* 1733 Paris; visited Italy 1754–65; otherwise worked Paris, where *d.* 1808.

Rosa, Salvator. Italian landscape and figure painter and etcher; *b.* 1615 Naples; visited Rome 1635–36 and 1639; in Florence 1640–49, afterwards in Rome, where *d.* 1673.

Rubens, Sir Peter Paul. Flemish figure, portrait and landscape painter and draughtsman; *b.* 1577 Siegen (Westphalia) of Flemish parents; trained Antwerp; in Italy 1600–8, afterwards worked mainly Antwerp, where *d.* 1640; visited Madrid (1603, 1628–9), Paris (1625), Holland (1624), London (1629–30).

Ruisdael, Jacob. Dutch landscape painter; *b.* 1628/9 Haarlem, where worked until *c.* 1655; afterwards worked Amsterdam, where *d.* 1682.

Rysbrack, John Michael. Anglo-Flemish sculptor; *b.* 1694 Antwerp; arrived England by 1720, where worked and *d.* 1770.

Sacchi, Andrea. Italian figure painter; *b.* 1599 Rome; trained Bologna, afterwards worked Rome, where *d.* 1661.

Sanchez Cotán, Juan. Spanish still-life painter; *b.* 1561 near Toledo, where trained and worked until 1603, then gave up painting to become a Carthusian monk; *d.* Granada 1627.

Sanctis, Francesco de. Italian architect; *b.* 1693 Rome, where worked and *d.* 1740.

Sassoferrato (properly Giovanni Battista Salvi). Italian figure painter; *b.* 1609 Sassoferrato; moved young to Rome, where studied under Domenichino; *d.* Rome 1685.

Serpotta, Giacomo. Italian sculptor; *b.* 1656 Palermo (Sicily), where worked and *d.* 1732.

Shaftesbury, Anthony Ashley Cooper, 3rd Earl of. English philosopher and writer on artistic theory; *b.* 1671 London; toured Europe, including Italy *c.* 1687–90; visited Holland 1699 and 1703–4; lived mainly in country in England; went to Italy again 1711, and *d.* Naples 1713.

Solimena, Francesco. Italian figure painter; *b.* 1657 Canale di Serino (Avellino); but settled Naples when young and worked and *d.* there 1747.

Soufflot, Jacques-Germain. French architect; *b.* 1713 Irancy (Yonne); visited Rome 1734–8 and 1749–51, with Cochin and Marigny; afterwards worked Paris and Lyon; *d.* Paris 1780.

Sprimont, Nicholas. Anglo-Flemish porcelain manufacturer; *b. c.* 1716 Flanders; in London from 1742 where manager (1749–58) and proprietor (1758–69) of Chelsea porcelain works; *d.* London 1771.

Steen, Jan. Dutch *genre* painter; *b.* 1626 Leyden; worked The Hague, Delft, Haarlem, Leyden, where *d.* 1679.

Stuart, James. Scottish architect, called 'Athenian Stuart'; *b.* 1713 London; with Revett, visited Naples 1748 and Athens 1751; otherwise mainly worked London, where *d.* 1788.

Stubbs, George. English animal painter; *b.* 1724 Liverpool; visited Italy 1754; settled London *c.* 1759, where *d.* 1806.

Terborch (or Ter Borch), **Gerard.** Dutch *genre* and portrait painter; *b.* 1617 Zwolle; visited Amsterdam and Haarlem (1632–4), London (1635); Rome (1640), possibly Madrid and Münster (Westphalia), 1646–8; afterwards worked Zwolle, where *d.* 1681.

Tiepolo, Giambattista (or Giovanni Battista). Italian figure painter; *b.* 1696 Venice, where trained and mainly worked; visited Würzburg, 1750–3 and Madrid, 1762, where *d.* 1770.

Tocqué, Louis. French portrait painter, pupil and son-in-law of Nattier; *b.* 1696 Paris, where mainly worked and *d.* 1772; visited St Petersburg and Copenhagen, 1756–9, and Copenhagen again 1769.

Tomé, Narciso. Spanish sculptor and decorator; active Valladolid and Toledo from 1715– *c.* 1740.

Tuby, Jean-Baptiste. French sculptor; *b.* 1635 Rome, but came to Paris young; worked Versailles (under Lebrun) and Paris, where *d.* 1700.

Vanbrugh, Sir John. English playwright and architect; *b.* 1664 London; began career as soldier; in France (partly in prison for spying) 1690–2; turned to architecture 1699 and worked throughout England; *d.* London 1726.

Van de Velde, Willem, the Younger. Dutch marine painter, son and pupil of Willem van de Velde the Elder, also marine artist; *b.* 1633 Leyden; family in Amsterdam from 1636 until father and son settled England 1671/2; *d.* London 1707.

Van Dyck, Sir Anthony. Flemish portrait and figure painter, draughtsman and etcher; *b.* 1599 Antwerp; pupil and assistant of Rubens; visited England, 1620, Italy (1621–5/6); settled England 1632, where *d.* 1641.

Van Dyck, Floris Claesz. Dutch still-life painter; *b.* 1575 probably Haarlem; in Italy *c.* 1600–10, then Haarlem, where *d.* 1651.

Van Goyen, Jan. Dutch landscape painter; *b.* 1596 Leyden; worked Haarlem *c.* 1617–34, then The Hague, where *d.* 1656.

Vanvitelli, Luigi. Italian architect and painter; *b.* 1700 Naples, trained Rome by father, Gaspar van Wittel, view painter; worked various parts of Italy but from 1751 chiefly in and near Naples, where *d.* 1773.

Vassé, François-Antoine. French decorative sculptor; *b.* 1681 Toulon; possibly arrived Paris 1698; working at Versailles from 1708; *d.* Paris 1736.

Velasquez, Diego Rodriguez de Silva y. Spanish portrait and figure painter; *b.* 1599 Seville; settled Madrid 1623, where *d.* 1660; visited Italy 1629–31 and 1648–51.

Vermeer, Jan. Dutch *genre* painter; *b.* 1632 Delft, where worked and *d.* 1675.

Vien, Joseph-Marie. French figure painter; *b.* 1716 Montpellier; in Rome 1743–50; afterwards Paris, where *d.* 1809; friend of Caylus and teacher of David.

Vittone, Bernardo Antonio. Italian architect; *b. c.* 1705 Turin; studied Rome when young, returning Turin where mainly worked and *d.* 1770.

Vouet, Simon. French figure painter; *b.* 1590 Paris; in Rome 1613–27; afterwards worked Paris, where *d.* 1649.

Wailly, Charles de. French architect; *b.* 1729 Paris; studied Rome, 1754–6; afterwards worked mainly Paris, where *d.* 1798.

Watteau, Antoine. French figure painter and draughtsman; *b.* 1684 Valenciennes; settled Paris 1702, where worked and *d.* 1721; visited England 1719–20.

Wedgwood, Josiah. English pottery manufacturer; *b.* 1730 Burslem (Staffs.); apprenticed as potter and started own business 1759 at Burslem; set up new factory at 'Etruria' 1773, where *d.* 1795.

West, Benjamin. American figure painter; *b.* 1739 Pennsylvania; visited Italy 1760–3, when settled London and *d.* there 1820; succeeded Reynolds in 1792 as President of the Royal Academy.

Willems, Joseph. Flemish porcelain modeller; *b. c.* 1690/1705 Brussels; in London (Chelsea), *c.* 1755–65; otherwise mainly worked Tournai, where *d.* 1766.

Wilson, Richard. English landscape painter; *b.* 1713/14 Penegoes (Montgomeryshire); trained London, then visited Italy 1750–7; afterwards lived mainly London; *d.* Penegoes 1782.

Wilton, Joseph. English sculptor; *b.* 1722 London; studied Belgium and Paris 1744–7 and Italy 1747–55; afterwards worked mainly London, where *d.* 1803.

Winckelmann, Johann Joachim. German archeologist and writer on classical art; *b.* 1717 Steindall (Brandenburg); studied theology at Halle, then lived mainly Dresden until 1755, when settled Rome; visited Vienna 1768, murdered in Trieste on way back to Rome the same year.

Wren, Sir Christopher. English architect and mathematician; *b.* 1632 Wiltshire; educated Oxford and London; visited Paris 1665; otherwise worked mainly Oxford, Cambridge and London, where *d.* 1723.

Zimmermann, Dominikus. German architect and stucco-worker; *b.* 1685 Wessobrunn (Bavaria); worked chiefly Bavaria, also Swabia and Franconia, often in collaboration with brother; *d.* Wies 1766.

Zimmermann, Johann Baptiste. German painter and stucco-worker, brother of the preceding; *b.* 1680 near Wessobrunn; from 1721 worked for Court at Augsburg; also worked Bavaria, Swabia and Franconia in collaboration with brother; *d.* Munich 1758.

Zurbaran, Francisco de. Spanish painter; *b.* 1598 near Badajoz; settled Seville 1629, where mainly worked; visited Madrid 1634, moved there 1658 and *d.* there 1664.

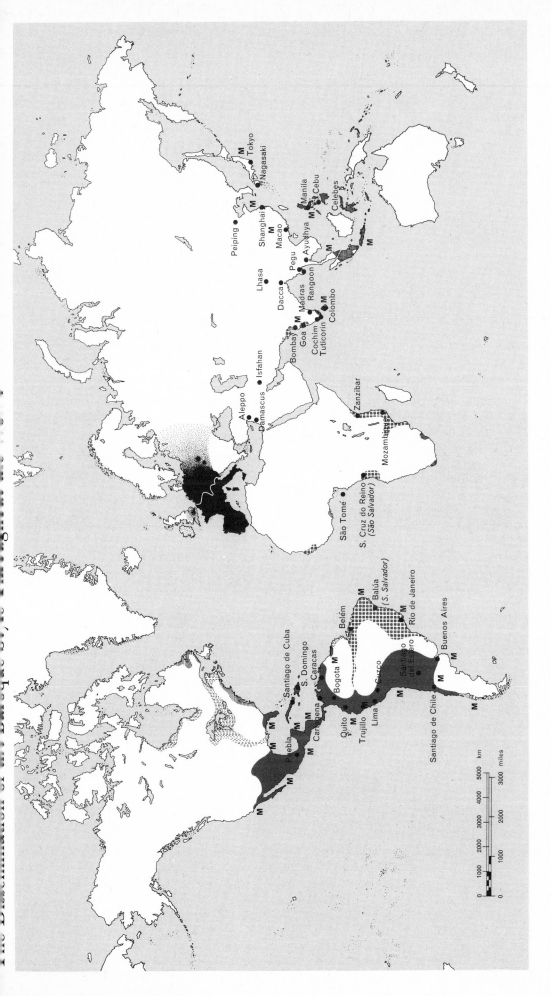

Peiping

Lhasa

Isfahan

Aleppo
Damascus

Shanghai
Macao

Pegu

Dacca
Madras
Bombay
Goa
Cochim
Tuticorin
Colombo

Tokyo
Nagasaki

Manila
Cebu
Celebes

Ayudhya

Rangoon

Zanzibar

Mozambique

São Tomé

S. Cruz do Reino
(São Salvador)

Santiago de Cuba
S. Domingo
Caracas
Bogota
Cartagena
Quito
Trujillo
Lima
Cuzco
Puebla
Santiago de Chile
Santiago del Estero
Buenos Aires

Belém
Balúa
(S. Salvador)
Rio de Janeiro

Varying impact of
Baroque in Europe. (The
white line shows the
western boundary of the
Holy Roman Empire).

Missionary settlements are identified by the names
of cities in which they had their principal seats,
and by the letter M (areas of maximum activity).

France

Holland

England

Spain

Portugal

km

0   1000   2000   3000   4000   5000

miles

0   1000   2000   3000

Further Reading List

GENERAL

Bazin, G. *Baroque and Rococo*, 1964 (Thames and Hudson)

Blunt, A. F. *Art and Architecture in France, 1500–1700*, 2nd ed. 1958 (Penguin Books, Pelican History of Art)

Fastnedge, R. *English Furniture Styles*, 1955 (Penguin)

Gerson, H. and Ter Kuile, E. H. *Art and Architecture in Belgium, 1600–1800*, (Penguin Books, Pelican History of Art)

De Goncourt, E. and J. *L'Art du dix-huitième siècle*, English ed. 1948 (Phaidon)

Haskell, F. *Patrons and Painters*, 1963 (Chatto and Windus)

Hautecoeur, L. *Histoire de l'architecture classique en France*, 1943–57 (Paris, A. & J. Picard)

Hawley, H. *Neo-Classicism: Style and Motif*, 1964 (Abrams for Cleveland Museum of Art)

Hempel, E. *Baroque Art and Architecture in Central Europe*, 1965 (Penguin Books, Pelican History of Art)

Holt, E. *A Documentary History of Art*, Vol. II, 1958 (Doubleday Anchor Books)

Kimball, Fiske *The Creation of the Rococo*, 1943 (Philadelphia Museum of Art) reprint 1965 (Norton Library, New York)

Kubler, G. and Soria, M. *Art and Architecture in Spain and Portugal and their American Dominions, 1500–1800*, 1954 (Penguin Books, Pelican History of Art)

Levey, M. *Painting in XVIIIth Century Venice*, 1959 (Phaidon)

Pope-Hennessy, J. *Italian High Renaissance and Baroque Sculpture*, 1963 (Phaidon)

Rosenberg, J. and Slive, S. *Dutch Art and Architecture, 1600–1800*, 1966 (Penguin Books, Pelican History of Art)

Schönberger, A. and Soehner, H. *The Age of Rococo*, 1960 (Thames and Hudson; McGraw-Hill)

Summerson, J. *Architecture in Britain, 1530–1830*, 4th, revised edition, 1963 (Penguin Books, Pelican History of Art)

Tapié, V.-L. *The Age of Grandeur*, 1960 (Weidenfeld and Nicolson)

Waterhouse, E. K. *Painting in Britain, 1530–1790*, 1953 (Penguin Books, Pelican History of Art)

Waterhouse, E. K. *Italian Baroque Painting*, 1962 (Phaidon)

Watson, F. J. B. *Wallace Collection Catalogues: Furniture*, 1956 (Wallace Collection)

Whinney, M. D. and Millar, O. *English Art, 1625–1714*, 1957 (Oxford History of English Art)

Whinney, M. D. *English Sculpture, 1530–1830*, 1964 (Penguin Books, Pelican History of Art)

Wittkower, R. *Art and Architecture in Italy, 1600–1750*, 1958 (Penguin Books, Pelican History of Art)

Wölfflin, H. *Renaissance and Baroque*, 1st German ed., 1888; 1964 (Collins, Fontana Library)

MONOGRAPHS

Antal, F. *Hogarth and his Place in European Art*, 1962 (Routledge)

Benesch, O. *The Drawings of Rembrandt*, 1954–7 (Phaidon)

Blunt, A. F. *Nicolas Poussin*, 1966 (Bollingen Foundation and Phaidon)

Briganti, G. *Pietro da Cortona*, 1962 (Florence, Sansoni)

Brookner, A. *Watteau*, 1966 (Paul Hamlyn)

Burckhardt, J. *Recollections of Rubens*, 1st German ed., 1898 English translation, 1950 (Phaidon)

Friedlaender, W. *Caravaggio Studies*, 1955 (Princeton)

Glück, G. *Van Dyck, Des Meisters Gemälde*, 1931 (Klassiker der Kunst)

Goldscheider, L. *Vermeer*, 1958 (Phaidon)

Held, J. *Rubens, Selected Drawings*, 1959 (Phaidon)

Hogarth, W. *The Analysis of Beauty*, first published 1753, ed., J. Burke, 1955 (Oxford)

Lafuente Ferrari, E. *Velasquez*, 1943 (Phaidon)

Morassi, A. *Tiepolo*, 1955–62 (Phaidon)

Münz, L. *Rembrandt's Etchings*, 1952 (Phaidon)

Oldenbourg, R. *Rubens, Des Meisters Gemälde*, 1921 (Klassiker der Kunst)

Parks, R. O. (ed.) *Piranesi*, 1961 (exhibition catalogue, Smith College, Northampton, Mass.)

Paulson, G. *Hogarth's Graphic Works*, 1965 (Yale University Press)

Reynolds, J. *Discourses on Art (1769–90)*, ed. R. R. Wark, 1959 (Huntington Library, San Marino, California)

Rosenberg, J. *Rembrandt*, revised ed., 1964 (Phaidon)

Röthlisberger, M. *Claude Lorrain: The Paintings*, 1961 (Yale University Press)

Salerno, L. *Salvator Rosa*, 1963 (Milan, Club del Libro)

Waterhouse, E. K. *Thomas Gainsborough*, 1958 (Edward Hulton)

Wildenstein, G. *Chardin*, 2nd edition 1963 (Zürich)

Wildenstein, G. *Fragonard*, 1960 (Phaidon)

Winckelmann, J. J. *Gedanken über die Nachahmung der griechischen Werke in der Malerei und Bildhauerkunst*, 1st German ed., 1755 (Dresden). English translation 1765 (London)

Wittkower, R. *Gian Lorenzo Bernini*, 1955 (Phaidon)

Wittkower, R. *Drawings of the Carracci at Windsor Castle*, 1952 (Phaidon)

Index

The numbers in heavy type refer to colour plates; italic numbers refer to black and white illustrations.

Acknowledgements

Photographs were provided by the following:

Colour: Lala Aufsberg, Nuremburg 14, 81, 82; Emil Bauer, Nuremburg 17, 93, 94; Boymans Museum, Rotterdam 63; Cleveland Museum of Art, Ohio 64; A. C. Cooper Ltd.. London 26, 51, 71; Dulwich College. London 19, 41; A. Frequin, The Hague 77; Leo Gundermann, Würzburg 8; Giraudon, Paris 33; Michael Holford, London 1, 3, 4, 6, 9, 12, 13, 17, 21, 22, 27, 29, 30, 32, 34, 35, 36, 37, 39, 42, 45, 47, 48, 50, 52, 54, 55, 57, 58, 59, 60, 62, 65, 66, 68, 69, 70, 72, 73, 74, 79, 80, 83, 84, 85, 88, 89, 92, 96, 100; D. Hughes Gilbey, London 44, 61; Jacqueline Hyde, Paris 11; A. F. Kersting, London 16, 86, 87, 99; Joseph Klima Jr., Detroit 43; MAS, Barcelona 95; Metropolitan Museum of Art, New York 25; Erwin Meyer, Vienna 5, 15, 24, 49; Mac Millar, San Diego 75; National Gallery of Canada, Ottawa 56; National Gallery of Art, Washington D.C. 23, 28, 38, 67; National Museum, Stockholm 53; Ramos, Madrid 2; Rijksmuseum, Amsterdam 7, 78; Scala, Florence 31, 40, 46; Tom Scott, Edinburgh 10; Edwin Smith, London by courtesy of Weidenfeld & Nicolson 90; Victoria & Albert Museum, London 14, 97, 98; R. Viollet, Paris 91; Nico Zomer, Haarlem 76

Black and White: Alinari, Rome 12, 13, 29, 30, 39, 41, 100; Anderson, Rome 2, 10, 31, 37, 38, 52, 53, 56, 62; Archives Photographiques, Paris 31, 58, 19, 80; British Museum, London 49, 78; Alfred Brod, London 66; A. C. Cooper, Ltd., London 11; Chicago Art Institute 25; Country Life, London 95; Courtauld Institute, London 6, 7, 8, 14, 37, 47, 48, 50, 70, 73, 75, 76, 84, 86, 89, 90, 91, 93, 99; John R. Freeman, London 1, 15, 17, 33, 54; Gabinetto Fotografico Nazionale, Rome 18, 30, 46; Giraudon, Paris 55, 71, 98; Dr. F. Hepner, Potsdam 83; Hermitage Museum, Leningrad 24; Michael Holford, London 3, 82; A. F. Kersting, London 43; Kunsthistorisches Museum, Vienna 65; F. Lugt, The Hague 74; Erwin Meyer, Vienna 44; Foto Marburg 16, 42; Mansell Collection, London 12, 94; MAS, Barcelona 61; Ministry of Works, London 57; National Gallery, London 67; National Museum, Berlin-Dahlem 51; Nottingham Castle Art Museum 97; Prado, Madrid 64; Presse-Bild, Munich 28; Rijksmuseum, Amsterdam 22; R.I.B.A., London 96; Edwin Smith, London, by courtesy of Weidenfeld and Nicholson 40; Service Photographique, Versailles 20; Victoria & Albert Museum, London 79, 88; Villani, Naples 45; Wallace Collection, London 23, 68, 69

pocket inside

1580 1590 1600 1610 1620 1630 1640 1650 1660 1670

Italy

L. Carracci (b. 1555)
Annibale Carracci (b 1560)
Caravaggio (b. 1573)
Gaulli
Guarini
C. Fontana
Maderna (b.1556)
Pozzo
Reni (b. 1575)
Guercino
Cortona
Sacchi
Algardi
Bernini
Borromini

Flanders

Rubens (b. 1577)
Van Dyck
Duquesnoy
Brouwer

Holland

Hals
Willem Van de Velde II
Van Goyen
Rembrandt
W. C. Heda
Terborch
Koninck
Cuyp
De Hooch
Steen
Ruisdael
Vermeer
Kalf

Spain

Montáñez (b. 1568)
Ribera
Zurbaran
Velasquez
Murillo

France

Vouet
G. De La Tour
Poussin
Lemercier
F. Mansart
Claude Lorrain
P. De Champaigne
Dughet
L. Le Nain
Lebrun
Puget
Coy
J. H. Mans
P. L

England

Jones (b. 1573)
Wren

46

Central Europe

heimer (b. 1578)

1580 1590 1600 1610 1620 1630 1640 1650 1660 1670

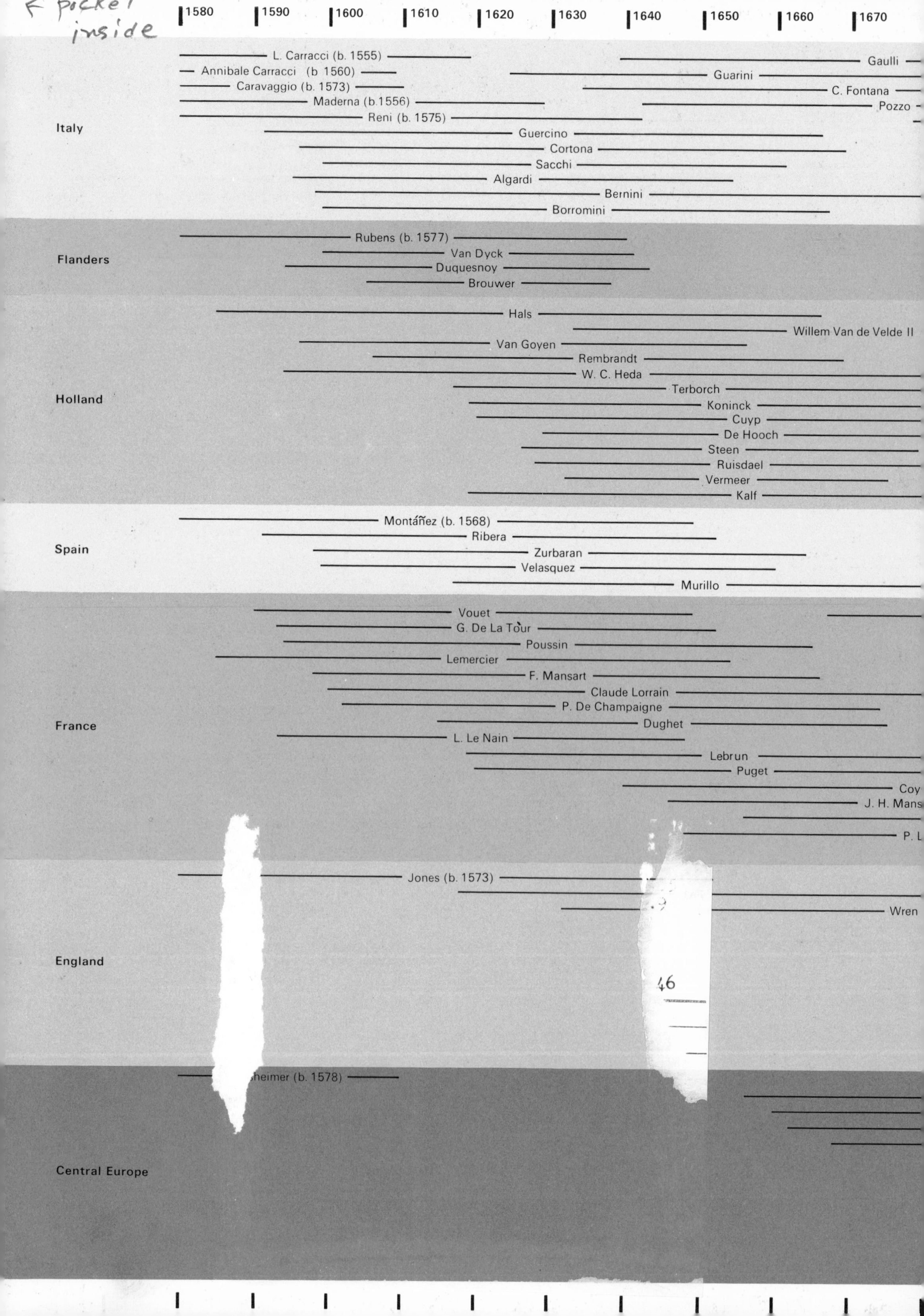